Statistics for Biology and Health

Series Editors

Mitchell Gail, Division of Cancer Epidemiology and Genetics, National Cancer Institute, Rockville, MD, USA

Jonathan M. Samet, Department of Epidemiology, School of Public Health, Johns Hopkins University, Baltimore, MD, USA

Statistics for Biology and Health (SBH) includes monographs and advanced textbooks on statistical topics relating to biostatistics, epidemiology, biology, and ecology.

More information about this series at http://www.springer.com/series/2848

Hans-Michael Kaltenbach

Statistical Design and Analysis of Biological Experiments

 Springer

Hans-Michael Kaltenbach
Department of Biosystems Science
and Engineering, ETH Zürich
Basel, Switzerland

ISSN 1431-8776   ISSN 2197-5671   (electronic)
Statistics for Biology and Health
ISBN 978-3-030-69643-6   ISBN 978-3-030-69641-2   (eBook)
https://doi.org/10.1007/978-3-030-69641-2

This Springer imprint is published by the registered company Springer Nature Switzerland AG
The registered company address is: Gewerbestrasse 11, 6330 Cham, Switzerland

For Elke

Preface

This book serves as an introduction to the design and analysis of experiments and uses examples from biology and life sciences. It is aimed at students and researchers wanting to gain or refresh knowledge of experimental design. Previous exposure to a standard statistics introduction would be helpful, but the required material is also reviewed in the first chapters to make the book self-contained. Most calculations are demonstrated in R, but should be easily transferable to other software.

The main feature of this book is the use of Hasse diagrams to construct and visualize a design. These diagrams clarify and simplify many ideas in experimental design yet have received little attention for teaching. They allow me to focus on the logical structure of an experimental design, facilitate comparisons between designs, cleanly separate treatments from units, and make the randomization more explicit. The visualization by Hasse diagrams encourages trying different modifications of a design, and an appropriate linear model and its specification in R are then easily derived from a given diagram. The analysis of variance techniques are carefully developed from basic principles, and I exploit Hasse diagrams to derive model specifications for linear mixed models as a modern alternative for more complex designs.

I aimed at keeping the book concise enough to be read cover-to-cover, yet to include the standard experimental designs found in biological research. I also discuss fractional factorial and response surface designs that proved invaluable for optimizing experimental conditions in my own practice. I believe that power analysis is an important part of experimental design and discuss this topic in detail; this includes 'portable power' as a quick-and-dirty tool for approximate sample size calculations that I have not seen anywhere else. Finally, I strongly emphasize estimation and effect sizes over testing and p-values and therefore discuss linear contrasts at considerable length; I also present standardized effect sizes usually not discussed in texts in the biomedical sciences.

In advancing through the material, I rely on a single main example—drugs and diets and their effect on mice—and use artificial data for illustrating the analyses. This approach has the advantage that the scientific questions and the experimental details can be handled rather easily, which allows me to introduce new designs and analysis techniques in an already familiar setting. It also emphasizes how the same treatments and experimental material can be organized and combined in different ways, which results in different designs with very different properties. Real experiments always have their own idiosyncrasies, of course, and rarely oblige the nice and clean design found in a textbook. To allow for this fact, I discuss several real-life examples with all their deviations from the most elegant designs, failed observations, and alternative interpretations of the results; these mostly originate from consultation and collaboration with colleagues at my department.

The book is organized in three main parts: Chaps. 1–3 introduce the design of experiments and the associated vocabulary, and provide a brief but thorough review of statistical concepts. Readers familiar with the material of a standard introductory statistics class might read Chap. 1 and then skim through Chaps. 2–3 to absorb the notation and get acquainted with Hasse diagrams and the introductory examples. The main designs and analysis techniques are discussed in Chaps. 4–8. This material provides the core of an introductory design of experiments class and includes completely randomized and blocked designs, factorial treatment designs, and split-unit designs. Chapters 9–10 introduce two more advanced methods and discuss the main ideas of fractional factorial designs for handling larger experiments and factor screening, and of response surface methods for optimization.

I am indebted to many people who contributed to this book. Lekshmi Dharmarajan and Julia Deichmann taught tutorials and often suffered through last-minute changes while the material and exposition took shape. They also provided valuable comments on earlier versions of the text. Many students at the Department of Biosystems Science and Engineering at ETH Zurich provided corrections and helpful feedback. Christian Lohasz and Andreas Hierlemann were kind enough to allow me to use their tumor diameter data, and Tania Roberts and Fabian Rudolf did the same for the yeast medium example. Cristina Loureiro Casalderrey worked on the yeast transformation example. My wife endured countless hours of me locked away in an office, staring and mumbling at a computer screen. Jörg Stelling generously granted me time and support for writing this book. Thank you all!

Finally, https://gitlab.com/csb.ethz/doe-book provides datasets and R-code for most examples as well as errata.

Basel, Switzerland   Hans-Michael Kaltenbach
November 2020

Contents

Chapter 1
Principles of Experimental Design

1.1 Introduction

The validity of conclusions drawn from a statistical analysis crucially hinges on the
manner in which the data are acquired, and even the most sophisticated analysis
will not rescue a flawed experiment. Planning an experiment and thinking about
the details of data acquisition are so important for a successful analysis that R. A.
Fisher—who single-handedly invented many of the experimental design techniques
we are about to discuss—famously wrote

> *To call in the statistician after the experiment is done may be no more than asking him to
> perform a* **post-mortem** *examination: he may be able to say what the experiment died of.*
> (Fisher 1938)

The *(statistical) design of experiments* provides the principles and methods for planning experiments and tailoring the data acquisition to an intended analysis. The
design and analysis of an experiment are best considered as two aspects of the same
enterprise: the goals of the analysis strongly inform an appropriate design, and the
implemented design determines the possible analyses.

The primary aim of designing experiments is to ensure that valid statistical and
scientific conclusions can be drawn that withstand the scrutiny of a determined skep-
tic. A good experimental design also considers that resources are used efficiently
and that estimates are sufficiently precise and hypothesis tests adequately powered.
It protects our conclusions by excluding alternative interpretations or rendering them
implausible. Three main pillars of experimental design are *randomization*, *replication*, and *blocking*, and we will flesh out their effects on the subsequent analysis as
well as their implementation in an experimental design.

An experimental design is always tailored toward pre-defined (primary) analyses,
and an efficient analysis and unambiguous interpretation of the experimental data is
often straightforward from a good design. This does not prevent us from conduct-

© Springer Nature Switzerland AG 2021
H.-M. Kaltenbach, *Statistical Design and Analysis of Biological Experiments*,
Statistics for Biology and Health, https://doi.org/10.1007/978-3-030-69641-2_1

ing additional analyses of interesting observations after the data are acquired, but these analyses can be subjected to more severe criticisms, and conclusions are more tentative.

In this chapter, we provide the wider context for using experiments in a larger research enterprise and informally introduce the main statistical ideas of experimental design. We use a comparison of two samples as our main example to study how design choices affect an analysis, but postpone a formal quantitative analysis to the next chapters.

1.2 A Cautionary Tale

For illustrating some of the issues arising in the interplay of experimental design and analysis, we consider a simple example. We are interested in comparing the enzyme levels measured in processed blood samples from laboratory mice, when the sample processing is done either with a kit from vendor A or a kit from a competitor B. For this, we take 20 mice and randomly select 10 of them for sample preparation with kit A, while the blood samples of the remaining 10 mice are prepared with kit B. The experiment is illustrated in Fig. 1.1A and the resulting data are given in Table 1.1.

One option for comparing the two kits is to look at the difference in average enzyme levels, and we find an average level of 10.32 for vendor A and 10.66 for vendor B. We would like to interpret their difference of -0.34 as the difference due to the two preparation kits and conclude whether the two kits give equal results or if

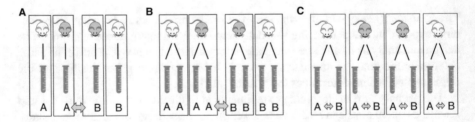

Fig. 1.1 Three designs to determine the difference between two preparation kits A and B based on four mice. **A** One sample per mouse. Comparison between averages of samples with the same kit. **B** Two samples per mouse treated with the same kit. Comparison between averages of mice with the same kit requires averaging responses for each mouse first. **C** Two samples per mouse each treated with a different kit. Comparison between two samples of each mouse, with differences averaged

Table 1.1 Measured enzyme levels from samples of 20 mice. Samples of 10 mice each were processed using a kit of vendors A and B, respectively

| A | 8.96 | 8.95 | 11.37 | 12.63 | 11.38 | 8.36 | 6.87 | 12.35 | 10.32 | 11.99 |
| B | 12.68 | 11.37 | 12.00 | 9.81 | 10.35 | 11.76 | 9.01 | 10.83 | 8.76 | 9.99 |

measurements based on one kit are systematically different from those based on the other kit.

Such interpretation, however, is only valid if the two groups of mice and their measurements are identical in all aspects except the sample preparation kit. If we use one strain of mice for kit A and another strain for kit B, any difference might also be attributed to inherent differences between the strains. Similarly, if the measurements using kit B were conducted much later than those using kit A, any observed difference might be attributed to changes in, e.g., mice selected, batches of chemicals used, device calibration, and any number of other influences. None of these competing explanations for an observed difference can be excluded from the given data alone, but a good experimental design allows us to render them (almost) arbitrarily implausible.

The second aspect for our analysis is the inherent uncertainty in our calculated difference: if we repeat the experiment, the observed difference will change each time, and this will be more pronounced for a smaller number of mice, among others. If we do not use a sufficient number of mice in our experiment, the uncertainty associated with the observed difference might be too large, such that random fluctuations become a plausible explanation for the observed difference. Systematic differences between the two kits, of practically relevant magnitude in either direction, might then be compatible with the data, and we can draw no reliable conclusions from our experiment.

In each case, the statistical analysis—no matter how clever—was doomed before the experiment was even started, while simple ideas from statistical design of experiments would have provided correct and robust results with interpretable conclusions.

1.3 The Language of Experimental Design

By an *experiment*, we understand an investigation where the researcher has full control over selecting and altering the experimental conditions of interest, and we only consider investigations of this type. The selected experimental conditions are called *treatments*. An experiment is *comparative* if the responses to several treatments are to be compared or contrasted. The *experimental units* are the smallest subdivision of the experimental material to which a treatment can be assigned. All experimental units given the same treatment constitute a *treatment group*. Especially in biology, we often compare treatments to a *control group* to which some standard experimental conditions are applied; a typical example is using a placebo for the control group and different drugs for the other treatment groups.

The values observed are called *responses* and are measured on the *response units*; these are often identical to the experimental units but need not be. Multiple experimental units are sometimes combined into *groupings* or *blocks*, such as mice grouped by litter or samples grouped by batches of chemicals used for their preparation. More generally, we call any grouping of the experimental material (even with group size one) a *unit*.

In our example, we selected the mice, used a single sample per mouse, deliberately chose the two specific vendors, and had full control over which kit to assign to which mouse. In other words, the two kits are the treatments and the mice are the experimental units. We took the measured enzyme level of a single sample from a mouse as our response, and samples are therefore the response units. The resulting experiment is comparative because we contrast the enzyme levels between the two treatment groups.

In this example, we can coalesce experimental and response units, because we have a single response per mouse and cannot distinguish a sample from a mouse in the analysis, as illustrated in Fig. 1.1A for four mice. Responses from mice with the same kit are averaged, and the kit difference is the difference between these two averages.

By contrast, if we take two samples per mouse and use the same kit for both samples, then the mice are still the experimental units, but each mouse now groups the two response units associated with it. Now, responses from the same mouse are first averaged, and these averages are used to calculate the difference between kits; even though eight measurements are available, this difference is still based on only four mice (Fig. 1.1B).

If we take two samples per mouse but apply each kit to one of the two samples, then the samples are both the experimental and response units, while the mice are blocks that group the samples. Now, we calculate the difference between kits for each mouse, and then average these differences (Fig. 1.1C).

If we only use one kit and determine the average enzyme level, then this investigation is still an experiment but is not comparative.

To summarize, the *design of an experiment* determines the *logical structure of the experiment*; it consists of (i) a set of treatments (the two kits), (ii) a specification of the experimental units (such as animals, cell lines, or samples) (the mice in Fig. 1.1A, B and the samples in Fig. 1.1C), (iii) a procedure for assigning treatments to units, and (iv) a specification of the response units and the quantity to be measured as a response (the samples and associated enzyme levels).

1.4   Experiment Validity

Before we embark on the more technical aspects of experimental design, we discuss three components for evaluating an experiment's validity: *construct validity*, *internal validity*, and *external validity*. These criteria are well-established in areas such as educational and psychological research, and have more recently been discussed for animal research (Würbel 2017) where experiments are increasingly scrutinized for their scientific rationale and their design and intended analyses.

1.4.1 Construct Validity

Construct validity concerns the choice of the experimental system for answering our research question. Is the system even capable of providing a relevant answer to the question?

Studying the mechanisms of a particular disease, for example, might require a careful choice of an appropriate animal model that shows a disease phenotype and is accessible to experimental interventions. If the animal model is a proxy for drug development for humans, biological mechanisms must be sufficiently similar between animal and human physiologies.

Another important aspect of the construct is the quantity that we intend to measure (the *measurand*) and its relation to the quantity or property we are interested in. For example, we might measure the concentration of the same chemical compound once in a blood sample and once in a highly purified sample, and these constitute two different measurands, whose values might not be comparable. Often, the quantity of interest (e.g., liver function) is not directly measurable (or even quantifiable) and we measure a *biomarker* instead. For example, pre-clinical and clinical investigations may use concentrations of proteins or counts of specific cell types from blood samples, such as the CD4+ cell count used as a biomarker for an immune system function.

1.4.2 Internal Validity

The internal validity of an experiment concerns the soundness of the scientific rationale, statistical properties such as precision of estimates, and the measures taken against the risk of bias. It refers to the validity of claims within the context of the experiment. The statistical design of experiments plays a prominent role in ensuring internal validity, and we briefly discuss the main ideas before providing the technical details and an application to our example in the subsequent sections.

Scientific Rationale and Research Question

The scientific rationale of a study is (usually) not immediately a statistical question. Translating a scientific question into a quantitative comparison amenable to statistical analysis is no small task and often requires careful consideration. It is a substantial, if non-statistical, benefit of using an experimental design that we are forced to formulate a precise-enough research question and decide on the main analyses required for answering it *before* we conduct the experiment. For example, the question: *is there a difference between placebo and drug?* is insufficiently precise for planning a statistical analysis and determine an adequate experimental design. What exactly is the drug treatment? What should the drug's concentration be and how is it administered? How do we make sure that the placebo group is comparable to the drug group in all other aspects? What do we measure and what do we mean by 'difference'? A shift in average response, a fold-change, change in response before and after treatment?

The scientific rationale also enters the choice of a potential control group to which we compare responses. The quote

The deep, fundamental question in statistical analysis is 'Compared to what?' (Tufte 1997)

highlights the importance of this choice.

There are almost never enough resources to answer all relevant scientific questions. We therefore concentrate on a few questions of the highest interest for our design, and the main purpose of the experiment is to answer these questions in the *primary analysis*. This intended analysis drives the experimental design to ensure relevant estimates can be calculated and have sufficient precision, and tests are adequately powered. This does not preclude us from conducting additional *secondary analyses* and *exploratory analyses*, but we are not willing to enlarge the experiment to ensure that strong conclusions can also be drawn from these analyses.

Risk of Bias

Experimental bias is a systematic difference in response between experimental units in addition to the difference caused by the treatments. The experimental units in the different groups are then not equal in all aspects other than the treatment applied to them. We saw several examples in Sect. 1.2.

Minimizing the risk of bias is crucial for internal validity and we look at some common measures to eliminate or reduce different types of bias in Sect. 1.5.

Precision and Effect Size

Another aspect of internal validity is the precision of estimates and the expected effect sizes. Is the experimental setup, in principle, able to detect a difference of relevant magnitude? Experimental design offers several methods for answering this question based on the expected heterogeneity of samples, the measurement error, and other sources of variation: *power analysis* is a technique for determining the number of samples required to reliably detect a relevant effect size and provide estimates of sufficient precision. More samples yield more precision and more power, but we have to be careful that *replication* is done at the right level: simply measuring a biological sample multiple times as in Fig. 1.1B yields more measured values, but is pseudo-replication for analyses. Replication should also ensure that the statistical uncertainties of estimates can be gauged from the data of the experiment itself, without additional untestable assumptions. Finally, the technique of *blocking*, shown in Fig. 1.1C, can remove a substantial proportion of the variation and thereby increase power and precision if we find a way to apply it.

1.4.3 External Validity

The external validity of an experiment concerns its *replicability* and the *generalizability* of inferences. An experiment is *replicable* if its results can be confirmed by an independent new experiment, preferably by a different lab and researcher. Experimental conditions in the replicate experiment usually differ from the original experiment, which provides evidence that the observed effects are robust to such changes. A much weaker condition on an experiment is *reproducibility*, the property that an independent researcher draws equivalent conclusions based on the data from this particular experiment, using the same analysis techniques. Reproducibility requires publishing the raw data, details on the experimental protocol, and a description of the statistical analyses, preferably with accompanying source code. Many scientific journals subscribe to *reporting guidelines* to ensure reproducibility, and these are also helpful for planning an experiment.

An important threat to replicability and generalizability are too tightly controlled experimental conditions, when inferences only hold for a specific lab under the very specific conditions of the original experiment. Introducing systematic heterogeneity and using multi-center studies effectively broadens the experimental conditions and therefore the inferences for which internal validity is available.

For *systematic heterogeneity*, experimental conditions are systematically altered in addition to the treatments, and treatment differences are estimated for each condition. For example, we might split the experimental material into several batches and use a different day of analysis, sample preparation, batch of buffer, measurement device, and lab technician for each batch. A more general inference is then possible if effect size, effect direction, and precision are comparable between the batches, indicating that the treatment differences are stable over the different conditions.

In *multi-center experiments*, the same experiment is conducted in several different labs and the results compared and merged. Multi-center approaches are very common in clinical trials and often necessary to reach the required number of patient enrollments.

Generalizability of randomized controlled trials in medicine and animal studies can suffer from overly restrictive eligibility criteria. In clinical trials, patients are often included or excluded based on co-medications and co-morbidities, and the resulting sample of eligible patients might no longer be representative of the patient population. For example, Travers et al. (2007) used the eligibility criteria of 17 random controlled trials of asthma treatments and found that out of 749 patients, only a median of 6% (45 patients) would be eligible for an asthma-related randomized controlled trial. This puts a question mark on the relevance of the trials' findings for asthma patients in general.

1.5   Reducing the Risk of Bias

1.5.1   *Randomization of Treatment Allocation*

If systematic differences other than the treatment exist between our treatment groups, then the effect of the treatment is *confounded* with these other differences and our estimates of treatment effects might be biased.

We remove such unwanted systematic differences from our treatment comparisons by randomizing the allocation of treatments to experimental units. In a *completely randomized design*, each experimental unit has the same chance of being subjected to any of the treatments, and any differences between the experimental units other than the treatments are distributed over the treatment groups. Importantly, randomization is the only method that also protects our experiment against *unknown* sources of bias: we do not need to know all or even any of the potential differences and yet their impact is eliminated from the treatment comparisons by random treatment allocation.

Randomization has two effects: (i) differences unrelated to treatment become part of the 'statistical noise' rendering the treatment groups more similar; and (ii) the systematic differences are thereby eliminated as sources of bias from the treatment comparison.

Randomization transforms systematic variation into random variation.

In our example, proper randomization would select 10 out of our 20 mice fully at random, such that the probability of any one mouse being picked is 1/20. These 10 mice are then assigned to kit A and the remaining mice to kit B. This allocation is entirely independent of the treatments and any properties of the mice.

To ensure random treatment allocation, some kind of random process needs to be employed. This can be as simple as shuffling a pack of 10 red and 10 black cards or using a software-based random number generator. Randomization is slightly more difficult if the number of experimental units is not known at the start of the experiment, such as when patients are recruited for an ongoing clinical trial (sometimes called *rolling recruitment*), and we want to have a reasonable balance between the treatment groups at each stage of the trial.

Seemingly random assignments 'by hand' are usually no less complicated than fully random assignments, but are always inferior. If surprising results ensue from the experiment, such assignments are subject to unanswerable criticism and suspicion of unwanted bias. Even worse are systematic allocations; they can only remove bias from known causes, and immediately raise red flags under the slightest scrutiny.

The Problem of Undesired Assignments

Even with a fully random treatment allocation procedure, we might end up with an undesirable allocation. For our example, the treatment group of kit A might—just by chance—contain mice that are all bigger or more active than those in the other treatment group. Statistical orthodoxy recommends using the design nevertheless, because only full randomization guarantees valid estimates of residual variance and unbiased estimates of effects. This argument, however, concerns the long-run properties of the procedure and seems of little help in this specific situation. Why should we care if the randomization yields correct estimates over many replications of the experiment, if the particular experiment is jeopardized?

Another solution is to create a list of all acceptable allocations and randomly choose one of these allocations for our experiment. The analysis should then reflect this restriction in the possible randomizations, which often renders this approach difficult to implement.

The most pragmatic method is to reject highly undesirable designs and compute a new randomization (Cox 1958). Undesirable allocations are unlikely to arise for large sample sizes, and we might accept a small bias in estimation for small sample sizes, when uncertainty in the estimated treatment effect is already high. In this approach, whenever we reject a particular outcome, we must also be willing to reject the outcome if we permute the treatment level labels. If we reject eight big and two small mice for kit A, then we must also reject two big and eight small mice. We must also be transparent and report a rejected allocation, so that critics may come to their own conclusions about potential biases and their remedies.

1.5.2 Blinding

Bias in treatment comparisons is also introduced if treatment allocation is random, but responses cannot be measured entirely objectively or if knowledge of the assigned treatment affects the response. In clinical trials, for example, patients might react differently when they know to be on a placebo treatment, an effect known as *cognitive bias*. In animal experiments, caretakers might report more abnormal behavior for animals on a more severe treatment. Cognitive bias can be eliminated by concealing the treatment allocation from technicians or participants of a clinical trial, a technique called *single blinding*.

If response measures are partially based on professional judgement (such as a clinical scale), the patient or the physician might unconsciously report lower scores for the placebo treatment, a phenomenon known as *observer bias*. Its removal requires *double blinding*, where treatment allocations are additionally concealed from the experimentalist.

Blinding requires randomized treatment allocation to begin with and substantial effort might be needed to implement it. Drug companies, for example, have to go to great lengths to ensure that a placebo looks, tastes, and feels similar enough to the actual drug. Additionally, blinding is often done by coding the treatment conditions and samples, and effect sizes and statistical significance are calculated before the code is revealed.

In clinical trials, double blinding creates a conflict of interest. The attending physicians do not know which patient received which treatment, and thus a potential accumulation of side-effects cannot be linked to a specific treatment. For this reason, clinical trials have a data monitoring committee, not involved in the final analysis, that performs intermediate analyses of efficacy and safety at pre-defined intervals. If severe problems are detected, the committee might recommend altering or aborting the trial. The same might happen if one treatment already shows overwhelming evidence of superiority and it becomes unethical to withhold this treatment from the other patients.

1.5.3 Analysis Plan and Registration

An often overlooked source of bias has been termed the *researcher degrees of freedom* or *garden of forking paths* in the data analysis. For any set of data, there are many different options for its analysis: some results might be considered outliers and be discarded, assumptions are made on error distributions and appropriate test statistics, and different covariates might be included into a regression model. Often, multiple hypotheses are investigated and tested, and analyses are done separately on various (overlapping) subgroups. Hypotheses formed after looking at the data require additional care in their interpretation; almost never will *p*-values for these *ad hoc* or *post hoc* hypotheses be statistically justifiable. Many different measured response variables invite *fishing expeditions*, where patterns in the data are sought without an underlying hypothesis. Only reporting those sub-analyses that gave 'interesting' findings invariably leads to biased conclusions and is called *cherry-picking* or *p-hacking* (or much less flattering names).

The statistical analysis is always part of a larger scientific argument, and we should consider the necessary computations in relation to building our scientific argument about the interpretation of the data. In addition to the statistical calculations, this interpretation requires substantial subject-matter knowledge and includes (many) non-statistical arguments. Two quotes highlight that experiment and analysis are a means to an end and not the end in itself.

> *There is a boundary in data interpretation beyond which formulas and quantitative decision procedures do not go, where judgment and style enter.* (Abelson 1995)

> *Often, perfectly reasonable people come to perfectly reasonable decisions or conclusions based on nonstatistical evidence. Statistical analysis is a tool with which we support reasoning. It is not a goal in itself.* (Bailar 1981)

There is often a gray area between exploiting researcher degrees of freedom to arrive at the desired conclusion and creative yet informed analyses of data. One way to navigate this area is to distinguish between *exploratory studies* and *confirmatory studies*. The former have no clearly stated scientific questions, but are used to generate interesting hypotheses by identifying potential associations or effects that are then further investigated. Conclusions from these studies are very tentative and must be reported honestly as such. In contrast, standards are much higher for confirmatory studies, which investigate a specific pre-defined scientific question. The analysis plan and pre-registration of an experiment are accepted means for demonstrating lack of bias due to researcher degrees of freedom, and separating primary from secondary analyses allows emphasizing the main goals of the study.

Analysis Plan

The *analysis plan* is written before conducting the experiment and details the measurands and estimands, the hypotheses to be tested together with a power and sample size calculation, a discussion of relevant effect sizes, detection and handling of outliers and missing data, as well as steps for data normalization such as transformations and baseline corrections. If a regression model is required, its factors and covariates are outlined. Particularly in biology, handling measurements below the limit of quantification and saturation effects require careful consideration.

In the context of clinical trials, the problem of *estimands* has become a recent focus of attention. An estimand is the target of a statistical estimation procedure, for example, the true average difference in enzyme levels between the two preparation kits. A common problem in many studies are *post-randomization events* that can change the estimand, even if the estimation procedure remains the same. For example, if kit B fails to produce usable samples for measurement in five out of 10 cases because the enzyme level was too low, while kit A could handle these enzyme levels perfectly fine, then this might severely exaggerate the observed difference between the two kits. Similar problems arise in drug trials, when some patients stop taking one of the drugs due to side-effects or other complications.

Registration

Registration of experiments is an even more severe measure used in conjunction with an analysis plan and is becoming standard in clinical trials. Here, information about the trial, including the analysis plan, procedure to recruit patients, and stopping criteria, are registered in a public database. Publications based on the trial then refer to this registration, such that reviewers and readers can compare what the researchers intended to do and what they actually did. Similar portals for pre-clinical and translational research are also available.

1.6 Notes and Summary

Notes

The problem of measurements and measurands is further discussed for statistics in Hand (1996) and specifically for biological experiments in Coxon et al. (2019). A general review of methods for handling missing data is Dong and Peng (2013). The different roles of randomization are emphasized in Cox (2009).

Two well-known reporting guidelines are the ARRIVE guidelines for animal research (Kilkenny et al. 2010) and the CONSORT guidelines for clinical trials (Moher et al. 2010). Guidelines describing the minimal information required for reproducing experimental results have been developed for many types of experimental techniques, including microarrays (MIAME), RNA sequencing (MINSEQE), metabolomics (MSI), and proteomics (MIAPE) experiments; the FAIRSHARE initiative provides a more comprehensive collection (Sansone et al. 2019).

The problems of experimental design in animal experiments and particularly translation research are discussed in Couzin-Frankel (2013). Multi-center studies are now considered for these investigations, and using a second laboratory already increases reproducibility substantially (Richter 2017; Richter et al. 2010; Voelkl et al. 2018; Karp 2018) and allows standardizing the treatment effects (Kafkafi et al. 2017). First attempts are reported of using designs similar to clinical trials (Llovera and Liesz 2016). Exploratory–confirmatory research and external validity for animal studies are discussed in Kimmelman et al. (2014) and Pound and Ritskes-Hoitinga (2018). Further information on pilot studies is found in Moore et al. (2011), Sim (2019), and Thabane et al. (2010).

The deliberate use of statistical analyses and their interpretation for supporting a larger argument was called *statistics as principled argument* (Abelson 1995). Employing useless statistical analysis without reference to the actual scientific question is *surrogate science* (Gigerenzer and Marewski 2014), and *adaptive thinking* is integral to meaningful statistical analysis (Gigerenzer 2002).

Summary

In an experiment, the investigator has full control over the experimental conditions applied to the experiment material. The experimental design gives the logical structure of an experiment: the units describing the organization of the experimental material, the treatments and their allocation to units, and the response. The statistical design of experiments includes techniques to ensure the internal validity of an experiment and methods to make the inference from experimental data efficient.

References

Abelson, R. P. (1995). Statistics as Principled Argument. Psychology Press.

Bailar III, J. C. (1981). "Bailar's laws of data analysis". In: Clinical Pharmacology & Therapeutics 20.1, pp. 113–119.

Couzin-Frankel, J. (2013). "When mice mislead". In: Science 342.6161, pp. 922–925.

Cox, D. R. (1958). Planning of Experiments. Wiley-Blackwell.

Cox, D. R. (2009). "Randomization in the design of experiments". In: International Statistical Review 77, pp. 415–429.

Coxon, C. H., C. Longstaff, and C. Burns (2019). "Applying the science of measurement to biology: Why bother?" In: PLOS Biology 17.6, e3000338.

Dong, Y. and C. Y. J. Peng (2013). "Principled missing data methods for researchers". In: SpringerPlus 2.1, pp. 1–17.

Fisher, R. A. (1938). "Presidential Address to the First Indian Statistical Congress". In: Sankhya: The Indian Journal of Statistics 4, pp. 14–17.

Gigerenzer, G. (2002). Adaptive Thinking: Rationality in the Real World. Oxford Univ Press.

Gigerenzer, G. and J. N. Marewski (2014). "Surrogate Science: The Idol of a Universal Method for Scientific Inference". In: Journal of Management 41.2, pp. 421–440.

Hand, D. J. (1996). "Statistics and the theory of measurement". In: Journal of the Royal Statistical Society A 159.3, pp. 445–492.

Kafkafi, N. et al. (2017). "Addressing reproducibility in single-laboratory phenotyping experiments". In: Nature Methods 14.5, pp. 462–464.

Karp, N. A. (2018). "Reproducible preclinical research-Is embracing variability the answer?" In: PLOS Biology 16.3, e2005413.

Kilkenny, C. et al. (2010). "Improving Bioscience Research Reporting: The ARRIVE Guidelines for Reporting Animal Research". In: PLOS Biology 8.6, e1000412.

Kimmelman, J., J. S. Mogil, and U. Dirnagl (2014). "Distinguishing between Exploratory and Confirmatory Preclinical Research Will Improve Translation". In: PLOS Biology 12.5, e1001863.

Llovera, G. and A. Liesz (2016). "The next step in translational research: lessons learned from the first preclinical randomized controlled trial". In: Journal of Neurochemistry 139, pp. 271–279.

Moher D.and Hopewell, S. et al. (2010). "CONSORT 2010 Explanation and Elaboration: updated guidelines for reporting parallel group randomised trials". In: BMJ: British Medical Journal 340.

Moore, C. G. et al. (2011). "Recommendations for planning pilot studies in clinical and translational research." In: Clinical and Translational Science 4.5, pp. 332–337.

Pound, P. and M. Ritskes-Hoitinga (2018). "Is it possible to overcome issues of external validity in preclinical animal research? Why most animal models are bound to fail". In: Journal of Translational Medicine 16.1, p. 304.

Richter, S. H. (2017). "Systematic heterogenization for better reproducibility in animal experimentation". In: Lab Animal 46.9, pp. 343–349.

Richter, S. H. et al. (2010). "Systematic variation improves reproducibility of animal experiments". In: Nature Methods 7.3, pp. 167–168.

Sansone, S.-A. et al. (2019). "FAIRsharing as a community approach to standards, repositories and policies". In: Nature Biotechnology 37.4, pp. 358–367.

Sim, J. (2019). "Should treatment effects be estimated in pilot and feasibility studies?" In: Pilot and Feasibility Studies 5.107, e1–e7.

Thabane, L. et al. (2010). "A tutorial on pilot studies: the what, why and how". In: BMC Medical Research Methodology 10.1, p. 1.

Travers J.and Marsh, S. et al. (2007). "External validity of randomised controlled trials in asthma: To whom do the results of the trials apply?" In: Thorax 62.3, pp. 219–233.

Tufte, E. (1997). Visual Explanations: Images and Quantities, Evidence and Narrative. 1st. Graphics Press.

Voelkl, B. et al. (2018). "Reproducibility of preclinical animal research improves with heterogeneity of study samples". In: PLOS Biology 16.2, e2003693.

Würbel, H. (2017). "More than 3Rs: The importance of scientific validity for harm-benefit analysis of animal research". In: Lab Animal 46.4, pp. 164–166.

Chapter 2
Review of Statistical Concepts

2.1 Introduction

We briefly review some basic concepts in statistical inference for analyzing a given set of data. The material roughly covers a typical introductory course in statistics: describing variability, estimating parameters, and testing hypotheses in the context of normally distributed data. The focus on the normal distribution avoids the need for more advanced mathematical and computational machinery and allows us to concentrate on the design rather than complex analysis aspects of an experiment in later chapters.

2.2 Probability

2.2.1 Random Variables and Distributions

Even under well-controlled conditions, replicate measurements will deviate from each other to a certain degree. In biological experiments, for example, two main sources of this variation are the heterogeneity of the experimental material and measurement uncertainties. After accounting for potential systematic effects, we can assign probabilities to these deviations.

A *random variable* maps a random event to a number, for example, to a (random) deviation of an observed measurement from the underlying true value. One often denotes a random variable by a capital letter and a specific *realization* or *outcome* by the corresponding lower case letter. For example, we might be interested in the level of a specific liver enzyme in mouse serum. The data in Table 2.1 show again the measured levels of 10 randomly selected mice, where one sample was taken from each mouse, processed using the kit from vendor A, and the enzyme level was quantified using a well-established assay.

© Springer Nature Switzerland AG 2021
H.-M. Kaltenbach, *Statistical Design and Analysis of Biological Experiments*,
Statistics for Biology and Health, https://doi.org/10.1007/978-3-030-69641-2_2

Table 2.1 Measured enzyme levels of 10 randomly selected mice

1	2	3	4	5	6	7	8	9	10
8.96	8.95	11.37	12.63	11.38	8.36	6.87	12.35	10.32	11.99

We formally describe the observation for the ith mouse by the outcome y_i of its associated random variable Y_i. The *(probability) distribution* of a random variable Y gives the probability of observing an outcome within a given interval. The *probability density function (pdf)* $f_Y(y)$ and the *cumulative distribution function (cdf)* $F_Y(y)$ of Y both describe its distribution. The area under the pdf between two values a and b gives the probability that a realization of Y falls into the interval $[a, b]$:

$$\mathbb{P}(Y \in [a, b]) = \mathbb{P}(a \leq Y \leq b) = \int_a^b f_Y(y') \, dy' = F_Y(b) - F_Y(a) \, ,$$

while the cdf gives the probability that Y will take any value lower or equal to y:

$$F_Y(y) = \mathbb{P}(Y \leq y) = \int_{-\infty}^y f_Y(y') dy' \, .$$

It is often reasonable to assume a *normal* (or *Gaussian*) *distribution* for the data. This distribution has two parameters μ and σ^2, and its probability density function is

$$f_Y(y; \mu, \sigma^2) = \frac{1}{\sqrt{2\pi}\sigma} \cdot \exp\left(-\frac{1}{2} \cdot \left(\frac{y - \mu}{\sigma} \right)^2 \right) \, ,$$

which yields the famous bell-shaped curve symmetric around a peak at μ and with width determined by σ. In our example, we have normally distributed observations with $\mu = 10$ and $\sigma^2 = 2$, which corresponds to the probability density function shown in Fig. 2.1.

We write $Y \sim N(\mu, \sigma^2)$ to say that the random variable Y has a normal distribution with parameters μ and σ^2. The density is symmetric around μ, and thus the probability that we observe an enzyme level lower than 10 is $\mathbb{P}(Y \leq 10) = 0.5$ in our example. The probability that the observed level falls above $a = 12.7$ is about $1 - F_Y(a) = \int_a^\infty f_Y(y') dy' = 0.025$ or 2.5%. More details on normal distributions and their properties are given in Sect. 2.2.6.

2.2.2 Quantiles

We are often interested in the α-*quantile* q_α below which a realization falls with given probability α such that

$$\mathbb{P}(Y \leq q_\alpha) = \alpha \, .$$

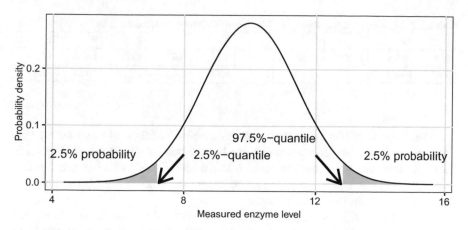

Fig. 2.1 Normal density for $N(\mu = 10, \ \sigma^2 = 2)$ with 2.5% and 97.5% quantiles (arrows) and areas corresponding to 2.5% probability under left and right tails (gray shaded areas)

The value q_α depends on the distribution of Y, and different distributions have different quantiles for the same α.

For our example, a new observation will be below the quantile $q_{0.025} = 7.23$ with probability 2.5% and below $q_{0.975} = 12.77$ with probability 97.5%. These two quantiles are indicated by arrows in Fig. 2.1, and each shaded area corresponds to a probability of 2.5%.

2.2.3 *Independence and Conditional Distributions*

The *joint distribution* of two random variables X and Y is denoted by

$$F_{X,Y}(x, y) = \mathbb{P}(X \le x, Y \le y) = \mathbb{P}(X \le x) \cdot \mathbb{P}(Y \le y | X \le x)$$

and is the probability that $X \le x$ and $Y \le y$ are simultaneously true. We can decompose it into the *marginal distribution* $\mathbb{P}(X \le x)$ and the *conditional distribution* $\mathbb{P}(Y \le y | X \le x)$. The conditional distribution is read as 'Y given X' and gives the probability that $Y \le y$ if we know that $X \le x$.

In our example, the enzyme levels Y_i and Y_j of two different mice are *independent*: their realizations y_i and y_j are both from the same distribution, but knowing the measured level of one mouse will tell us nothing about the level of the other mouse, hence $\mathbb{P}(Y_i \le y_i \,|\, Y_j \le y_j) = \mathbb{P}(Y_i \le y_i)$.

The joint probability of two independent variables is the product of the two individual probabilities. For example, the probability that both mouse i and mouse j yield measurements below 10 is $\mathbb{P}(Y_i \le 10, \ Y_j \le 10) = \mathbb{P}(Y_i \le 10 | Y_j \le 10) \cdot \mathbb{P}(Y_j \le 10) = \mathbb{P}(Y_i \le 10) \cdot \mathbb{P}(Y_j \le 10) = 0.5 \cdot 0.5 = 0.25$.

Table 2.2 Measured enzyme levels of 10 lab mice, each mouse measured twice

1	2	3	4	5	6	7	8	9	10
8.96	8.95	11.37	12.63	11.38	8.36	6.87	12.35	10.32	11.99
8.82	9.13	11.37	12.50	11.75	8.65	7.63	12.72	10.51	11.80

As an example of two dependent random variables, we might take a second sample from each mouse and measure the enzyme level of this sample, similar to Fig. 1.1B. The resulting data are shown in Table 2.2. We immediately observe that the first and second measurements from the same mouse are much more similar than any measurements from two different mice.

We denote by $Y_{i,1}$ and $Y_{i,2}$ the two measurements of the ith mouse. Then, $Y_{i,1}$ and $Y_{j,1}$ are independent since they are measurements of two different mice and $\mathbb{P}(Y_{i,1} \leq 10 | Y_{j,1} \leq 10) = \mathbb{P}(Y_{i,1} \leq 10) = 0.5$, for example. In contrast, $Y_{i,1}$ and $Y_{i,2}$ are not independent, because they are two measurements of the same mouse: knowing the result of the first measurement gives us a strong indication about the value of the second measurement and $\mathbb{P}(Y_{i,2} \leq 10 | Y_{i,1} \leq 10)$ is likely to be much higher than 0.5, because if the first sample yields a level below 10, the second will also tend to be below 10. We discuss this in more detail shortly when we look at covariances and correlations.

2.2.4 Expectation and Variance

Instead of working with the full distribution of a random variable Y, it is often sufficient to summarize its properties by the *expectation* and *variance*, which roughly speaking give the position around which the density function spreads and the dispersion of values around this position, respectively.

The *expected value* (or *expectation*, also called *mean* or *average*) of a random variable Y is a measure of its *location*. It is defined as the weighted average of all possible realizations y of Y, which we calculate by integrating over the values y and multiplying each value with the probability density $f_Y(y)$ of Y's distribution:

$$\mathbb{E}(Y) = \int_{-\infty}^{+\infty} y \cdot f_Y(y) \, dy \, .$$

The expectation is linear and the following arithmetic rules apply for any two random variables X and Y and any non-random constant a:

$$\mathbb{E}(a) = a \, , \;\; \mathbb{E}(a + X) = a + \mathbb{E}(X) \, , \;\; \mathbb{E}(a \cdot X) = a \cdot \mathbb{E}(X) \, , \;\; \mathbb{E}(X + Y) = \mathbb{E}(X) + \mathbb{E}(Y).$$

If X and Y are independent, then the expectation of their product is the product of the expectations:

$$\mathbb{E}(X \cdot Y) = \mathbb{E}(X) \cdot \mathbb{E}(Y) ,$$

but note that in general $\mathbb{E}(X / Y) \neq \mathbb{E}(X) / \mathbb{E}(Y)$ even for independent variables. The expectation is often denoted by μ.

The *variance* of a random variable Y, often denoted by σ^2, is defined as

$$\text{Var}(Y) = \mathbb{E}\left((Y - \mathbb{E}(Y))^2\right) = \mathbb{E}(Y^2) - \mathbb{E}(Y)^2 ,$$

the expected distance of a value of Y from its expectation, where the distance is measured as the squared difference. It is a measure of *dispersion*, describing how wide values spread around their expected value.

For a non-random constant a and two random variables X and Y, the following arithmetic rules apply for variances:

$$\text{Var}(a) = 0 , \ \ \text{Var}(a + Y) = \text{Var}(Y) , \ \ \text{Var}(a \cdot Y) = a^2 \cdot \text{Var}(Y) .$$

If X and Y are independent, then

$$\text{Var}(X + Y) = \text{Var}(X) + \text{Var}(Y) .$$

Moreover, the variance of the *difference* of two independent random variables is larger than each individual variance since

$$\text{Var}(X - Y) = \text{Var}(X + (-1) \cdot Y) = \text{Var}(X) + (-1)^2 \cdot \text{Var}(Y) = \text{Var}(X) + \text{Var}(Y).$$

For a normally distributed random variable $Y \sim N(\mu, \sigma^2)$, the expectation and variance completely specify the full distribution, since $\mu = \mathbb{E}(Y)$ and $\sigma^2 = \text{Var}(Y)$. For our example distribution in Fig. 2.1, the expectation $\mu = 10$ provides the location of the maximum of the density, and the variance $\sigma^2 = 2$ corresponds to the width of the density curve around this location. The relation between a distribution's parameters and its expectation and variance is less direct for many other distributions.

Given the expectation and variance of a random variable Y, we can define a new random variable Z with expectation zero and variance one by shifting and scaling:

$$Z = \frac{Y - \mathbb{E}(Y)}{\sqrt{\text{Var}(Y)}} \quad \text{has} \quad \mathbb{E}(Z) = 0 \text{ and } \text{Var}(Z) = 1 .$$

If Y is normally distributed with parameters μ and σ^2, then $Z = (Y - \mu)/\sigma$ is also normally distributed with parameters $\mu = 0$ and $\sigma^2 = 1$. In general, however, the distribution of Z will be of a different kind than the distribution of Y. For example, Y might be a concentration and take only positive values, but Z is centered around zero and will take positive and negative values.

The realizations of n independent and identically distributed random variables Y_i are often summarized by their arithmetic mean $\bar{Y} = \frac{1}{n} \sum_{i=1}^{n} Y_i$. Since $\bar{Y}$ is a function of random variables, it is itself a random variable and therefore has a distribution. Indeed, if the individual variables Y_i are normally distributed, then $\bar{Y}$ also has a normal distribution. Assuming that all Y_i have expectation μ and variance σ^2, we can determine the expectation and variance of the arithmetic mean. First, we find that its expectation is identical to the expectation of the individual Y_i:

$$\mathbb{E}(\bar{Y}) = \mathbb{E}\left(\frac{1}{n} \sum_{i=1}^{n} Y_i\right) = \frac{1}{n} \sum_{i=1}^{n} \mathbb{E}(Y_i) = \frac{1}{n} \sum_{i=1}^{n} \mu = \frac{1}{n} \cdot n \cdot \mu = \mu .$$

However, its variance is smaller than the individual variances, and decreases with the number of random variables:

$$\mathrm{Var}(\bar{Y}) = \mathrm{Var}\left(\frac{1}{n} \sum_{i=1}^{n} Y_i\right) = \frac{1}{n^2} \sum_{i=1}^{n} \mathrm{Var}(Y_i) = \frac{1}{n^2} \cdot n \cdot \sigma^2 = \frac{\sigma^2}{n} .$$

For our example, the average of the 10 measurements of enzyme levels has expectation $\mathbb{E}(\bar{Y}) = \mathbb{E}(Y_i) = 10$, but the variance $\mathrm{Var}(\bar{Y}) = \sigma^2/10 = 0.2$ is only one-tenth of the individual variances $\mathrm{Var}(Y_i) = \sigma^2 = 2$. The average of 10 measurements therefore shows less dispersion around the mean than each individual measurement. Since the sum of normally distributed random variables is again normally distributed, we thus know that $\bar{Y} \sim N(\mu, \sigma^2/n)$ with $n = 10$ for our example.

We often describe the *scale* of the distribution by the *standard deviation*:

$$\mathrm{sd}(Y) = \sqrt{\mathrm{Var}(Y)} ,$$

which is a measure of dispersion in the same unit as the random variable. While the variance of the sum of independent random variables is the sum of their variances, the variance does not behave nicely with changes in the measurement scale: if Y is measured in meters, its variance is given in square-meters. A shift to centimeters then multiplies all measurements by 100, but the variance by $100^2 = 10\,000$. In contrast, the standard deviation behaves nicely under changes in scale, but not under addition:

$$\mathrm{sd}(a \cdot Y) = |a| \cdot \mathrm{sd}(Y) \quad \text{but} \quad \mathrm{sd}(X \pm Y) = \sqrt{\mathrm{sd}(X)^2 + \mathrm{sd}(Y)^2} ,$$

for independent X and Y. For our example, $\mathrm{sd}(Y_i) = 1.41$ and $\mathrm{sd}(\bar{Y}) = 0.45$.

2.2.5 Covariance and Correlation

The joint dispersion of X and Y is described by their *covariance*

$$\text{Cov}(X, Y) = \mathbb{E}\big((X - \mathbb{E}(X)) \cdot (Y - \mathbb{E}(Y))\big) = \mathbb{E}(X \cdot Y) - \mathbb{E}(X) \cdot \mathbb{E}(Y) \,,$$

which measures the dependency between X and Y: the covariance is large and positive if a large deviation of X from its expectation is associated with a large deviation of Y in the same direction; it is large and negative if these directions are reverse.

The covariance of a random variable with itself is its variance $\text{Cov}(X, X) = \text{Var}(X)$, and the covariance is zero if X and Y are independent (the converse is not true!). The covariance is linear in both arguments, in particular:

$$\text{Cov}(a \cdot X + b \cdot Y, \; Z) = a \cdot \text{Cov}(X, \; Z) + b \cdot \text{Cov}(Y, \; Z) \,,$$

and similarly for the second argument. A related measure is the *correlation*

$$\text{Corr}(X, Y) = \frac{\text{Cov}(X, Y)}{\text{sd}(X) \cdot \text{sd}(Y)} \,,$$

which is a unitless number between -1 and $+1$.

The covariance is the missing ingredient to extend our formulas for the expectation and variance to the case of two dependent variables by

$$\mathbb{E}(X \cdot Y) = \mathbb{E}(X) \cdot \mathbb{E}(Y) + \text{Cov}(X, Y)$$

and

$$\text{Var}(X + Y) = \text{Var}(X) + \text{Var}(Y) + 2 \cdot \text{Cov}(X, Y)$$
$$\text{Var}(X - Y) = \text{Var}(X) + \text{Var}(Y) - 2 \cdot \text{Cov}(X, Y) \,,$$

which both reduce to the previous formulas if the variables are independent and $\text{Cov}(X, Y) = 0$.

In our first example, the measurements of enzyme levels in 10 mice are independent. Therefore, $\text{Cov}(Y_i, Y_i) = \text{Var}(Y_i)$ and $\text{Cov}(Y_i, Y_j) = 0$ for two different mice i and j.

In our second example, two samples were measured for each mouse. We can write the random variable $Y_{i,j}$ for the jth measurement of the ith mouse as the sum of a random variable M_i capturing the difference of the true enzyme level of mouse i to the expectation $\mu = 10$, and a random variable $S_{i,j}$ capturing the difference between the observed enzyme level $Y_{i,j}$ and the true level of mouse i. Then,

$$Y_{i,j} = \mu + M_i + S_{i,j} \,,$$

where $\mu = 10$, $M_i \sim N(0, \sigma_m^2)$, and $S_{i,j} \sim N(0, \sigma_e^2)$. The ith mouse then has average response $\mu + M_i$, and the first measurement deviates from this by $S_{i,1}$. If M_i and $S_{i,j}$ are all independent, then the overall variance σ^2 is decomposed into the two *variance components*: $\mathrm{Var}(Y_{i,j}) = \sigma^2 = \sigma_m^2 + \sigma_e^2 = \mathrm{Var}(M_i) + \mathrm{Var}(S_{i,j})$.

If we plot the two measurements $Y_{i,1}$ and $Y_{i,2}$ of each mouse in a scatterplot as in Fig. 2.2A, we notice the high correlation: whenever the first measurement is high, the second measurement is also high. The covariance between these variables is

$$
\begin{aligned}
&\mathrm{Cov}(Y_{i,1}, Y_{i,2}) \\
&= \mathrm{Cov}(\mu + M_i + S_{i,1}, \mu + M_i + S_{i,2}) \\
&= \mathrm{Cov}(M_i + S_{i,1}, M_i + S_{i,2}) \\
&= \underbrace{\mathrm{Cov}(M_i, M_i)}_{=\mathrm{Var}(M_i)} + \underbrace{\mathrm{Cov}(M_i, S_{i,2})}_{=0 \text{ (independence)}} + \underbrace{\mathrm{Cov}(S_{i,1}, M_i)}_{=0 \text{ (independence)}} + \underbrace{\mathrm{Cov}(S_{i,1}, S_{i,2})}_{=0 \text{ (independence)}} \\
&= \sigma_m^2,
\end{aligned}
$$

and the correlation is

$$
\mathrm{Corr}(Y_{i,1}, Y_{i,2}) = \frac{\mathrm{Cov}(Y_{i,1}, Y_{i,2})}{\mathrm{sd}(Y_{i,1}) \cdot \mathrm{sd}(Y_{i,2})} = \frac{\sigma_m^2}{\sigma_m^2 + \sigma_e^2}.
$$

With $\sigma_m^2 = 1.9$ and $\sigma_e^2 = 0.1$, the correlation is extremely high at 0.95. This can also be seen in Fig. 2.2B, which shows the two measurements for each mouse separately. Here, the magnitude of the differences between the individual mice (gray points) is measured by σ_m^2, while the magnitude of differences between the two measurements of any mouse (plus and cross) is much smaller and is given by σ_e^2.

As the third example, we consider the arithmetic mean of independent random variables and determine the covariance $\mathrm{Cov}(Y_i, \bar{Y})$. Because $\bar{Y}$ is computed using Y_i, these two random variables cannot be independent, and we expect a non-zero covariance. Using the linearity of the covariance, we get

$$
\mathrm{Cov}(Y_i, \bar{Y}) = \mathrm{Cov}\left(Y_i, \frac{1}{n}\sum_{j=1}^{n} Y_j\right) = \frac{1}{n}\sum_{j=1}^{n}\mathrm{Cov}(Y_i, Y_j) = \frac{1}{n}\mathrm{Cov}(Y_i, Y_i) = \frac{\sigma^2}{n},
$$

using the fact that $\mathrm{Cov}(Y_i, Y_j) = 0$ for $i \neq j$. Thus, the covariance decreases with increasing number of random variables, because the variation of the average $\bar{Y}$ depends less and less on each single variable Y_i.

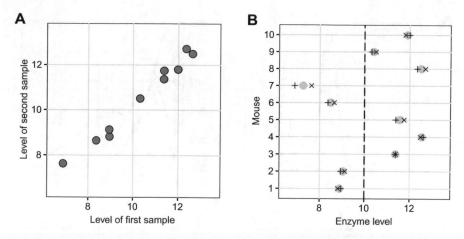

Fig. 2.2 **A** Scatterplot of the enzyme levels of the first and second samples for each mouse. The points lie close to a line, indicating a very high correlation between the two samples. **B** Average (gray) and first (plus) and second (cross) sample enzyme levels for each mouse. The dashed line is the average level in the mouse population

2.2.6 Some Important Distributions

We encounter four families of distributions in the next chapters, and briefly gather some of their properties here for future reference. We assume that all data follow some normal distribution describing the errors due to sampling and measurement, for example. Derived from that are (i) the χ^2-distribution related to estimates of variance, (ii) the F-distribution related to the ratio of two variance estimates, and (iii) the t-distribution related to estimates of means and differences in means.

Normal Distribution

The normal distribution is well-known for its bell-shaped density, shown in Fig. 2.3A for three combinations of its two parameters μ and σ^2. The special case of $\mu = 0$ and $\sigma^2 = 1$ is called the *standard normal distribution*.

The omnipresence of the normal distribution can be partly explained by the *central limit theorem*, which states that if we have a sequence of random variables with identical distribution (for example, describing the outcome of many measurements), then their average will have a normal distribution, no matter what distribution each single random variable has. Technically, if $Y_1, \ldots, Y_n$ are independent and identically distributed with mean $\mathbb{E}(Y_i) = \mu$ and variance $\mathrm{Var}(Y_i) = \sigma^2$, then the arithmetic mean $\bar{Y}_n = (Y_1 + \cdots + Y_n)/n$ approaches a normal distribution as n increases:

$$\bar{Y}_n \sim N(\mu, \sigma^2/n) \text{ as } n \to \infty.$$

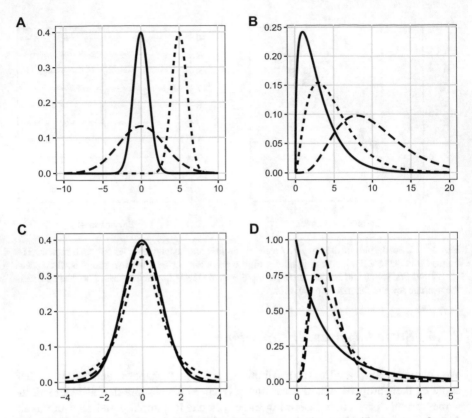

Fig. 2.3 Probability densities. **A** Normal distributions: standard normal with mean $\mu = 0$, standard deviation $\sigma = 1$ (solid); shifted: $\mu = 5$, $\sigma = 1$ (dotted); scaled: $\mu = 0$, $\sigma = 3$ (dashed). **B** χ^2-distributions with 3 (solid), 5 (dotted), and 10 (dashed) degrees of freedom. **C** t-distributions with 2 (dotted) and 10 (dashed) degrees of freedom and standard normal density (solid). **D** F-distributions with ($n = 2$, $m = 10$) numerator/denominator degrees of freedom (solid), respectively, ($n = 10$, $m = 10$) (dotted) and ($n = 10$, $m = 100$) (dashed)

If $X \sim N(\mu_X, \sigma_X^2)$ and $Y \sim N(\mu_Y, \sigma_Y^2)$ are two normally distributed random variables, then their sum $X + Y$ is also normally distributed with parameters

$$\mathbb{E}(X + Y) = \mu_X + \mu_Y \quad \text{and} \quad \text{Var}(X + Y) = \sigma_X^2 + \sigma_Y^2 + 2 \cdot \text{Cov}(X, Y) .$$

Any normally distributed variable $Y \sim N(\mu, \sigma^2)$ can be transformed into a standard normal variable by subtracting the mean and dividing by the standard deviation: $Z = \frac{Y-\mu}{\sigma} \sim N(0, 1)$. The α-quantile of the standard normal distribution is denoted by z_α, and $z_\alpha = -z_{1-\alpha}$ by symmetry.

A few properties of the normal distribution are helpful to remember. If $Y \sim N(\mu, \sigma^2)$, then the probability that Y will fall into an interval $\mu \pm \sigma$ of one standard deviation around the mean is about 70%; for two standard deviations $\mu \pm 2\sigma$, the probability is about 95%; and for three standard deviations, it is larger than 99%. In particular, the 2.5% and 97.5% quantiles of a standard normal distribution are $z_{0.025} = -1.96$ and $z_{0.975} = +1.96$ and can be conveniently taken to be ± 2 for approximate calculations.

χ^2-Distribution

If $Y_1, \ldots, Y_n \sim N(0, 1)$ are independent and standard normal random variables, then the sum of their squares has a χ^2 (read: 'chi-square') -distribution with n degrees of freedom:

$$\sum_{i=1}^{n} Y_i^2 \sim \chi_n^2 .$$

This distribution has expectation n and variance $2n$, and the degrees of freedom n is its only parameter. It occurs when estimating the variance of a normally distributed random variable from a set of measurements, and is used to establish confidence intervals of variance estimates (Sect. 2.3.4.2). We denote by $\chi_{\alpha, n}^2$ the α-quantile on n degrees of freedom.

Its densities for 3, 5, and 10 degrees of freedom are shown in Fig. 2.3B. Note that the densities are asymmetric, only defined for non-negative values, and the maximum is at $n - 2$ and does not coincide with the expected value (for $n = 5$ the expected value is 5, while the maximum is at 3).

The sum of two independent χ^2-distributed random variables is again χ^2-distributed, with degrees of freedom the sum of the two individual degrees of freedom:

$$X \sim \chi_n^2 \text{ and } Y \sim \chi_m^2 \implies X + Y \sim \chi_{n+m}^2 .$$

Moreover, if $\bar{Y} = \sum_{i=1}^{n} Y_i / n$ is the average of n standard normally distributed variables, then

$$\sum_{i=1}^{n} (Y_i - \bar{Y})^2 \sim \chi_{n-1}^2$$

also has a χ^2-distribution, where one degree of freedom is 'lost' because we can calculate any single summand from the remaining $n - 1$.

If the random variables $Y_i \sim N(\mu_i, 1)$ are normally distributed with unit variance but individual (potentially non-zero) means μ_i, then $\sum_i Y_i^2 \sim \chi^2(\lambda)$ has a *noncentral* χ^2-*distribution* with *noncentrality parameter* $\lambda = \sum_i \mu_i^2$. This distribution plays a role in sample size determination, for example.

t-Distribution

If X has a standard normal distribution, and Y is independent of X and has a χ_n^2-distribution, then the random variable

$$\frac{X}{\sqrt{Y/n}} \sim t_n$$

has a t-distribution with n degrees of freedom. This distribution has expectation zero and variance $n/(n-2)$ and occurs frequently when studying the sampling distribution of normally distributed random variables whose expectation and variance are estimated from the data. It most famously underlies the Student's t-test, which we discuss in Sect. 2.4.1, and plays a prominent role in finding confidence intervals of estimates of averages (Sect. 2.3.4.1). The α-quantile is denoted by $t_{\alpha,n} = -t_{1-\alpha,n}$.

The density of the t-distribution looks very similar to that of the standard normal distribution, but has 'thicker tails'. As the degrees of freedom n get larger, the two distributions become more similar, and the t-distribution approaches the standard normal distribution in the limit: $t_\infty = N(0,1)$. Two examples of t-distributions are shown in Fig. 2.3C together with the standard normal distribution for comparison.

If the numerator $X \sim N(\mu, 1)$ has a non-zero expectation, then the random variable $X/(\sqrt{Y/n}) \sim t_n(\eta)$ has a *noncentral t-distribution* with noncentrality parameter $\eta = \mu/(\sqrt{Y/n})$; this distribution is used in determining sample sizes for t-tests, for example.

F-Distribution

If $X \sim \chi_n^2$ and $Y \sim \chi_m^2$ are two independent χ^2-distributed random variables, then their ratio, scaled by the degrees of freedom, has an F-distribution with n numerator degrees of freedom and m denominator degrees of freedom, which are its two parameters:

$$\frac{X/n}{Y/m} \sim F_{n,m} \ .$$

This distribution has expectation $m/(m-2)$ and variance $2m^2(n+m-2)/(n(m-2)^2(m-4))$. It occurs when investigating the ratio of two variance estimates and plays a central role in the *analysis of variance*, a key method in experimental design (Chap. 4). An F-distribution is only defined for positive values and is asymmetric with a pronounced right tail (it has positive *skew*). The sum of two F-distributed random variables does *not* have an F-distribution. The α-quantile is denoted by $F_{\alpha,n,m} = 1/F_{1-\alpha,m,n}$ and Fig. 2.3D gives three examples of F-densities.

For $m = \infty$, the $F_{n,m}$-distribution corresponds to the χ_n^2-distribution $\chi_n^2 = F_{n,\infty}$, and for $n = 1$, the F-distribution corresponds to the square of a t-distribution $F_{1,m} = t_m^2$. If $X \sim \chi_n^2(\lambda)$ has a noncentral χ^2-distribution, then $(X/n)/(Y/m) \sim F_{n,m}(\lambda)$ has a *noncentral F-distribution with noncentrality parameter* λ; hence $\chi_n^2(\lambda) = F_{n,\infty}(\lambda)$ and $t_m^2(\eta) = F_{1,m}(\lambda = \eta^2)$.

2.3 Estimation

In practice, the expectation, variance, and other parameters of a distribution are often unknown and need to be determined from data such as the 10 measurements in Table 2.1. For the moment, we do not make assumptions on the distribution of these measured levels, and simply write $Y_i \sim (\mu, \sigma^2)$ to emphasize that the observations Y_i have (unknown) mean μ and (unknown) variance σ^2. These two parameters should give us an adequate picture of the expected enzyme level and the variation in the mouse population.

The enzyme level y_i of mouse i will deviate from the population average μ by a random amount e_i, and we can write it as

$$y_i = \mu + e_i ,$$

where the deviations $e_i \sim (0, \sigma^2)$ are distributed around a zero mean. There are two parts to this model: one part (μ) describes the *mean structure* of our problem, that is, the location of the distribution; the other (e_i and the associated variance σ^2) the *variance structure*, that is, the random deviations around the location. The deviations $e_i = y_i - \mu$ are called *residuals* and capture the *unexplained variation* with *residual variance* σ^2. The diagram in Fig. 2.4 gives a visual representation of this situation: reading from top to bottom, it shows the increasingly finer partition of the data. On top, the *factor* **M** corresponds to the population mean μ and gives the coarsest summary of the data. We write this factor in bold to indicate that its parameter is a fixed number. The summary is then refined by the next finer partition, which here already corresponds to 10 individual mice. To each mouse is associated the difference from its value (the observations y_i) and the next-coarser partition (the population mean μ), and the factor *(Mouse)* corresponds to the 10 residuals $e_i = \mu - y_i$. In contrast to the population mean, the residuals are random and will change in a replication of the experiment. We indicate this fact by writing the factor in italics and parentheses.

The number of parameters associated with each granularity is given as a super-script (one population mean and 10 residuals); the subscript gives the number of *independent* parameters. While there are 10 residuals, there are only nine *degrees*

Fig. 2.4 Experiment structure of enzyme levels from 10 randomly sampled mice

of freedom for their values, since knowing nine residuals and the population mean allows calculation of the value for the tenth residual. The degrees of freedom are easily calculated from the diagram: take the number of parameters (the superscript) of each factor and subtract the degrees of freedom of every factor above it.

Since *(Mouse)* subdivides the partition of **M** into a finer partition of the data, we say that *(Mouse)* is *nested* in **M**.

Our task is now threefold: (i) provide an *estimate* of the expected population enzyme level μ from the given data $y_1, \ldots, y_{10}$, (ii) provide an estimate of the population variance σ^2, and (iii) quantify the uncertainty of those estimates.

The *estimand* is a population parameter θ and an *estimator* of θ is a function that takes data $y_1, \ldots, y_n$ as an input and returns a number that is a 'good guess' of the true value of the parameter. There might be several sensible estimators for a given parameter, and statistical theory and assumptions on the data usually provide insight into which estimator is most appropriate.

We denote an estimator of a parameter θ by $\hat{\theta}$ and sometimes use $\hat{\theta}_n$ to emphasize its dependence on the sample size. The *estimate* is the value of $\hat{\theta}$ that results from a specific set of data. Standard statistical theory provides us with methods for constructing estimators, such as *least squares*, which requires few assumptions, and *maximum likelihood*, which requires postulating the full distribution of the data but can then better leverage them. Since the data are random, so is the estimate, and the estimator is therefore a random variable with an expectation and a variance.

2.3.1 Properties of Estimators

The *bias* of an estimator is the difference between its expectation and the true value of the parameter:

$$\text{bias}\left(\hat{\theta}\right) = \mathbb{E}\left(\hat{\theta}\right) - \theta = \mathbb{E}\left(\hat{\theta} - \theta\right) .$$

We call an estimator *unbiased* or *accurate* if $\text{bias}(\hat{\theta}) = 0$. Then, some realizations of our data will yield estimates that are larger than the true value, others will yield lower ones, but our estimator returns the correct parameter value on average. In contrast, a biased or inaccurate estimator systematically over- or underestimates the true parameter value (Fig. 2.5).

While it is often difficult to work out the distribution of an arbitrary estimator, the distribution of an estimator derived from the maximum likelihood principle tends to a normal distribution for large sample sizes such that

$$\frac{\hat{\theta} - \theta}{\text{sd}(\hat{\theta})} \sim N(0, 1) \quad \text{for n} \to \infty$$

for an unbiased maximum likelihood estimator.

An estimator is *consistent* if the estimates approach the true value of the parameter when increasing the sample size indefinitely.

2.3.2 Estimators of Expectation and Variance

We now consider estimators for the mean and variance, as well as covariance and correlations as concrete examples. These estimators form the basis for more complex estimation problems that we encounter later on. We look at properties like bias and consistency in more detail to illuminate these concepts with simple examples.

The most common estimator for the expectation μ is the arithmetic mean

$$\hat{\mu} = \bar{y} = \frac{1}{n} \sum_{i=1}^{n} y_i .$$

Our previous calculations show that

$$\mathbb{E}(\hat{\mu}) = \mathbb{E}\left(\frac{1}{n} \sum_{i=1}^{n} y_i\right) = \mu ,$$

and hence this estimator is unbiased.

From the data in Table 2.1, for example, we find the estimate $\hat{\mu} = 10.32$ for a parameter value of $\mu = 10$.

Similarly, the maximum likelihood estimator for the variance is

$$\dot{\sigma}^2 = \frac{1}{n} \sum_{i=1}^{n} (y_i - \mu)^2 ,$$

but we would need μ to calculate it. In order to make this estimator operational, we plug in the estimator $\hat{\mu}$ instead of μ, which gives the estimator

$$\hat{\sigma}^2 = \frac{1}{n} \sum_{i=1}^{n} (y_i - \hat{\mu})^2 .$$

This estimator is biased, as we easily verify with a direct if somewhat lengthy calculation.

$$\mathbb{E}(\hat{\sigma}^2) = \mathbb{E}\left(\frac{1}{n}\sum_{i=1}^{n}(y_i - \hat{\mu})^2\right) = \mathbb{E}\left(\frac{1}{n}\sum_{i=1}^{n}((y_i - \mu) - (\hat{\mu} - \mu))^2\right)$$

$$= \frac{1}{n}\sum_{i=1}^{n}\left(\mathbb{E}((y_i - \mu)^2) - 2\cdot\mathbb{E}\left((y_i - \mu)(\hat{\mu} - \mu)\right) + \mathbb{E}((\hat{\mu} - \mu)^2)\right)$$

$$= \frac{1}{n}\sum_{i=1}^{n}\left(\sigma^2 - 2\text{Cov}(y_i, \hat{\mu}) + \frac{\sigma^2}{n}\right) = \frac{1}{n}\sum_{i=1}^{n}\left(\sigma^2 - 2\frac{\sigma^2}{n} + \frac{\sigma^2}{n}\right)$$

$$= \left(\frac{n-1}{n}\right)\cdot\sigma^2 < \sigma^2 .$$

Hence, this estimator systematically underestimates the true variance. The bias decreases with increasing sample size, as $(n-1)/n$ approaches one for large n.

Moreover, the bias is known and can be explicitly calculated in advance, because it only depends on the sample size n and not on the data y_i. We can therefore remedy this bias simply by multiplying the estimator with $\frac{n}{n-1}$ and thus arrive at an unbiased estimator for the variance

$$\hat{\sigma}^2 = \frac{n}{n-1}\hat{\sigma}^2 = \frac{1}{n-1}\sum_{i=1}^{n}(y_i - \hat{\mu})^2 .$$

The resulting denominator $n-1$ corresponds exactly to the degrees of freedom for the residuals in the diagram (Fig. 2.4).

An estimator for the standard deviation σ is $\hat{\sigma} = \sqrt{\hat{\sigma}^2}$; this estimator is biased, and no unbiased estimator for σ exists.

For our example, we find that the biased estimator is $\hat{\sigma}^2 = 3.39$ and the unbiased estimator is $\hat{\sigma}^2 = 3.77$ (for $\sigma^2 = 2$); the estimate of the standard deviation is then $\hat{\sigma} = 1.94$ (for $\sigma = 1.41$).

With the same arguments, we find that

$$\widehat{\text{Cov}}(X, Y) = \frac{1}{n-1}\sum_{i=1}^{n}\left((x_i - \hat{\mu}_X)\cdot(y_i - \hat{\mu}_Y)\right)$$

is an estimator of the covariance $\text{Cov}(X, Y)$ of two random variables X and Y with means μ_X and μ_Y, based on n sampled pairs $(x_1, y_1), \ldots, (x_n, y_n)$.

Combining this with the estimates of the standard deviations yields an estimator for the correlation ρ of

$$\hat{\rho} = \frac{\widehat{\text{Cov}}(X, Y)}{\widehat{\text{sd}}(X)\cdot\widehat{\text{sd}}(Y)} .$$

For our two-sample example, we find a covariance between first and second samples of $\widehat{\text{Cov}}(Y_{i,1}, Y_{i,2}) = 3.47$ and thus a very high correlation of $\hat{\rho} = 0.99$.

2.3.3 Standard Error and Precision

Since an estimator $\hat{\theta}$ is a random variable, it has a variance $\mathrm{Var}(\hat{\theta})$, which quantifies the dispersion around its mean. Its standard deviation $\mathrm{sd}(\hat{\theta})$ is called the *standard error* of the estimator and denoted by $\mathrm{se}(\hat{\theta})$; its reciprocal $1/\mathrm{se}(\hat{\theta})$ is the *precision*.

The distribution of an estimator not only depends on the distribution of the data but also on the size n of the sample. Larger samples result in lower dispersion, and the standard error thus decreases with increasing sample size.

For example, the estimator $\hat{\mu}$ has variance $\mathrm{Var}(\hat{\mu}) = \sigma^2/n$ and its standard error

$$\mathrm{se}(\hat{\mu}) = \sigma/\sqrt{n}$$

decreases with rate $1/\sqrt{n}$ when increasing the sample size n. The estimator is therefore consistent and concentrates around the true value with increasing sample size. For doubling its precision, we require four times the amount of data, and to get one more decimal place precise, a 100-fold. We estimate the standard error by plugging in an estimate of σ^2:

$$\widehat{\mathrm{se}}(\hat{\mu}) = \hat{\sigma}/\sqrt{n} .$$

For our example, we find that $\widehat{\mathrm{se}}(\hat{\mu}) = 0.61$ and the standard error is only about 6% of the estimated expectation $\hat{\mu} = 10.32$, indicating high relative precision.

For normally distributed data, the ratio

$$\frac{\dot{\sigma}^2}{\sigma^2} = \frac{1}{n} \sum_{i=1}^{n} \left(\frac{y_i - \mu}{\sigma} \right)^2$$

has a χ^2-distribution. To compensate for the larger uncertainty when using an estimate $\hat{\mu}$ instead of the true value, we need to adjust the degrees of freedom and arrive at the distribution of the variance estimator $\hat{\sigma}^2$:

$$\frac{(n-1)\hat{\sigma}^2}{\sigma^2} \sim \chi^2_{n-1} .$$

This distribution has variance $2(n-1)$, and from it we find the variance of the estimator to be

$$2(n-1) = \mathrm{Var}\left(\frac{(n-1)\hat{\sigma}^2}{\sigma^2} \right) = \frac{(n-1)^2}{\sigma^4} \mathrm{Var}(\hat{\sigma}^2) \iff \mathrm{Var}(\hat{\sigma}^2) = \frac{2}{n-1}\sigma^4 .$$

Hence, $\mathrm{se}(\hat{\sigma}^2) = \sqrt{2/(n-1)}\sigma^2$, the estimator is consistent, and doubling the precision again requires about four times as much data. The standard error now depends on the true parameter value σ^2, and larger variances are more difficult to estimate precisely. For our example, we find that the variance estimate $\hat{\sigma}^2 = 3.77$ has an esti-

Fig. 2.5 Accuracy and
precision of an estimator

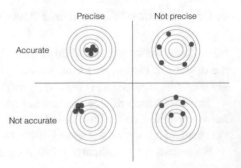

mated standard error of $\widehat{\mathrm{se}}(\hat{\sigma}^2) = 1.78$. This is an example of an accurate (unbiased) yet imprecise estimator.

Indeed, the precision and accuracy of an estimator are complementary properties as illustrated in Fig. 2.5: the precision describes the dispersion of estimates around the expected value while accuracy describes the systematic deviation between expectation and true value. Estimates might cluster very closely around a single value, indicating high precision, yet estimates might still be incorrect for a biased estimator when this single value is not the true parameter value.

2.3.4 Confidence Intervals

The estimators for expectation and variance are examples of *point estimators* and provide a single number as the 'best guess' of the true parameter from the data. The standard error quantifies the uncertainty of a point estimate: an estimate of the average enzyme level based on 100 mice is more precise than an estimate based on only two mice.

A *confidence interval* of a parameter θ is another way for quantifying the uncertainty that additionally takes account of the full distribution of the estimator. The interval contains all values of the parameter that are compatible with the observed data up to a specified degree. The $(1 - \alpha)$-confidence interval of an estimator $\hat{\theta}$ is an interval $[a(\hat{\theta}), b(\hat{\theta})]$ such that

$$\mathbb{P}\left(a(\hat{\theta}) \leq \theta \leq b(\hat{\theta})\right) = 1 - \alpha \,.$$

We call $a(\hat{\theta})$ and $b(\hat{\theta})$ the *lower* and *upper confidence limit*, respectively, and abbreviate them as LCL and UCL. The *confidence level* $1 - \alpha$ quantifies the degree of being 'compatible with the data'. The higher the confidence level (the lower α), the wider the interval, until a 100%-confidence interval includes all possible values of the parameter and becomes useless.

While not strictly required, we always choose the confidence limits a and b such that the left and right tails each cover half of the required confidence level. This provides the shortest possible confidence interval.

The confidence interval equation is a probability statement about a *random interval covering* θ and *not* a statement about the probability that θ is contained in a given interval. For example, it is incorrect to say that, having computed a 95%-confidence interval of $[-2, 2]$, the true population parameter has a 95% probability of being larger than -2 and smaller than $+2$. Such a statement would be nonsensical, because any given interval either contains the true value (which is a fixed number) or it does not, and there is no probability attached to this.

One correct interpretation is that the proportion of intervals containing the correct value θ is $(1 - \alpha)$ under repeated sampling and estimation. This interpretation is helpful in contexts like quality control, where the 'same' experiment is done repeatedly and there is thus a direct interest in the proportion of intervals containing the true value. For most biological experiments, we do not anticipate repeating them over and over again. Here, an equivalent interpretation is that our specific confidence interval, computed from the data of our experiment, has a $(1 - \alpha)$ probability to contain the correct parameter value.

For computing a confidence interval, we need to derive the distribution of the estimator; for maximum likelihood estimators, this distribution is normal for large sample sizes, and

$$\frac{\theta - \hat{\theta}}{\mathrm{se}(\hat{\theta})} \sim N(0, 1) \,,$$

and therefore

$$z_{\alpha/2} \leq \frac{\theta - \hat{\theta}}{\mathrm{se}(\hat{\theta})} \leq z_{1-\alpha/2} \quad \text{with probability} \quad 1 - \alpha \,,$$

where z_α is the α-quantile of the standard normal distribution. Rearranging terms yields the well-known asymptotic confidence interval for an unbiased maximum likelihood estimator

$$\left[\hat{\theta} + z_{\alpha/2} \cdot \mathrm{se}(\hat{\theta}), \ \hat{\theta} + z_{1-\alpha/2} \cdot \mathrm{se}(\hat{\theta}) \right] \,.$$

For some standard estimators, the exact distribution might not be normal but is nevertheless known. If the distribution is unknown or difficult to compute, we can use computational methods such as bootstrapping to find an approximate confidence interval.

2.3.4.1   Confidence Interval for Mean Estimate

For normally distributed data $y_i \sim N(\mu, \sigma^2)$, we know that

$$\frac{\hat{\mu} - \mu}{\sigma/\sqrt{n}} \sim N(0, 1) \quad \text{and} \quad \frac{\hat{\mu} - \mu}{\hat{\sigma}/\sqrt{n}} \sim t_{n-1} \, .$$

The normal distribution is appropriate if the variance is known or if the sample size n is large.

We then construct a $(1 - \alpha)$-confidence interval based on the normal distribution:

$$\left[\hat{\mu} + z_{\alpha/2} \cdot \sigma/\sqrt{n}, \ \hat{\mu} + z_{1-\alpha/2} \cdot \sigma/\sqrt{n} \, \right] = \hat{\mu} \pm z_{\alpha/2} \cdot \sigma/\sqrt{n} \, .$$

The equality holds because $z_{\alpha/2} = -z_{1-\alpha/2}$. For small sample sizes n, we need to take account of the additional uncertainty from replacing σ by $\hat{\sigma}$. The exact confidence interval of μ is then

$$\hat{\mu} \pm t_{\alpha/2,\, n-1} \cdot \widehat{se}(\hat{\mu}) = \hat{\mu} \pm t_{\alpha/2,\, n-1} \cdot \hat{\sigma}/\sqrt{n} \, .$$

For our data, the 95%-confidence interval based on the normal approximation is [9.11, 11.52] and the interval based on the t-distribution [8.93, 11.71] is wider. For larger sample sizes n, the difference between the two intervals quickly becomes negligible.

To further illustrate confidence intervals, we return to our enzyme level example. The true $N(10, 2)$-density of enzyme levels is shown in Fig. 2.6A (solid line) and the $N(10, 2/10)$-density of the estimator of the expectation based on 10 samples is shown as a dotted line. The rows of Fig. 2.6B are the enzyme levels of 10 replicates of 10 randomly sampled mice as gray points, with the true average $\mu = 10$ shown

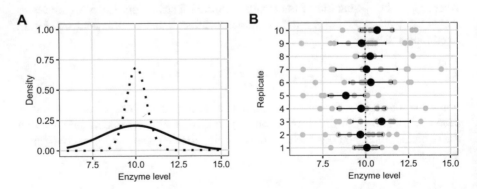

Fig. 2.6   A Distribution of enzyme levels in population (solid) and of average (dotted); **B** Levels measured for 10 replicates of 10 randomly sampled mice each (gray points) with estimated mean for each replicate (black points) and their 95%-confidence intervals (black lines) compared to true mean (dotted line)

as a vertical dotted line. The resulting estimates of the expectation and their 95%-confidence intervals are given as black points and lines, respectively. The 10 estimates vary much less than the actual measurements and fall both above and below the true expectation. Since the confidence intervals are based on the random samples, they have different lower and upper confidence limits and different lengths. For a confidence limit of 95%, we expect that 1 out of 20 confidence intervals does not cover the true value on average; an example is the interval of replicate five that covers only values below the true expectation.

2.3.4.2 Confidence Interval for Variance Estimate

For the variance estimator, recall that $(n - 1)\hat{\sigma}^2/\sigma^2$ has a χ^2-distribution. It follows that

$$\chi^2_{\alpha/2,\, n-1} \leq \frac{(n-1)\hat{\sigma}^2}{\sigma^2} \leq \chi^2_{1-\alpha/2,\, n-1} \quad \text{with probability } 1\text{-}\alpha \,,$$

where $\chi^2_{\alpha,\, n}$ is the α-quantile of a χ^2-distribution with n degrees of freedom. Rearranging terms, we find the $(1 - \alpha)$-confidence interval

$$\frac{(n-1)\hat{\sigma}^2}{\chi^2_{1-\alpha/2,\, n-1}} \leq \sigma^2 \leq \frac{(n-1)\hat{\sigma}^2}{\chi^2_{\alpha/2,\, n-1}} \,.$$

This interval does not simplify to the form $\hat{\theta} \pm q_\alpha \cdot$ se, because the χ^2-distribution is not symmetric and $\chi^2_{\alpha,\, n} \neq -\chi^2_{1-\alpha,\, n}$.

For our example, we estimate the population variance as $\hat{\sigma}^2 = 3.77$ (for a true value of $\sigma^2 = 2$). With $n - 1 = 9$ degrees of freedom, the two quantiles for a 95%-confidence interval are $\chi^2_{0.025,\, 9} = 2.7$ and $\chi^2_{0.975,\, 9} = 19.02$, leading to an interval of $[1.78, 12.56]$. The interval covers the true value, but its width indicates that our variance estimate is quite imprecise. A confidence interval for the standard deviation σ is calculated by taking square-roots of the lower and upper confidence limits of the variance. This yields a 95%-confidence interval of $[1.34, 3.54]$ for our estimate $\hat{\sigma} = 1.94$ (true value of $\sigma = 1.41$).

2.3.5 Estimation for Comparing Two Samples

Commonly, we are interested in comparing properties of two different samples to determine if there are detectable differences between their distributions. For example, if we always used a sample preparation kit of vendor A for our enzyme level measurements, but vendor B now also has a (maybe cheaper) kit available, we may want to establish if the two kits yield similar results. For this example, we are specifically interested in establishing if there is a systematic difference between measurements

Table 2.3 Measured enzyme levels for 20 mice, measured by kits from two vendors A and B

A	8.96	8.95	11.37	12.63	11.38	8.36	6.87	12.35	10.32	11.99
B	12.68	11.37	12.00	9.81	10.35	11.76	9.01	10.83	8.76	9.99

Fig. 2.7 Experiment for estimating the difference in enzyme levels by randomly assigning 10 mice each to two vendors

based on kit A compared to kit B or if measurements from one kit are more dispersed than from the other.

We denote by μ_A, μ_B the expectations and by σ_A^2, σ_B^2 the variances of the enzyme levels $y_{i,A}$ and $y_{i,B}$ measured by kit A and B, respectively. Our interest focuses on two *effect sizes*: we measure the systematic difference between the kits by the difference $\Delta = \mu_A - \mu_B$ of their expected values, and we measure the difference in variances by their proportion $\Pi = \sigma_A^2/\sigma_B^2$. Since we already have 10 measurements with our standard kit A, our proposed experiment is to select another 10 mice and measure their levels using kit B. This results in the data shown in Table 2.3, whose first row is identical to our previous data.

Our goal is to establish an estimate of the difference in means and the proportion of variances, and to calculate confidence intervals for these estimates.

The logical structure of the experiment is shown in the diagram in Fig. 2.7.

The data from this experiment are described by the *grand mean* $\mu = (\mu_A + \mu_B)/2$, given by the factor **M**; this is the 'best guess' for the value of any datum y_{ij} if nothing else is known. If we know which vendor was assigned to the datum, a better 'guess' is the corresponding mean μ_A or μ_B. The factor **Vendor** is associated with the two differences $\mu_A - \mu$ and $\mu_B - \mu$. Since we can calculate μ_A, say, from μ and μ_B, the degrees of freedom for this factor are one. Finally, the next finer partition of the data is into the individual observations; their $2 \cdot 10$ residuals $e_{ij} = y_{ij} - \mu_i$ are associated with the factor *(Mouse)*, and only 18 of the residuals are independent given the two group means. In this diagram, *(Mouse)* is nested in **Vendor** and **Vendor** is nested in **M** since each factor further subdivides the partition of the factor above. This implies that *(Mouse)* is also nested in **M**.

The diagram corresponds to the model

$$y_{ij} = \mu + \delta_i + e_{ij}$$

for the data, where $\delta_i = \mu_i - \mu$ are the deviations from grand mean to group mean and $\mu_A = \mu + \delta_A$ and $\mu_B = \mu + \delta_B$. The three parameters μ, δ_A, and δ_B are unknown but fixed quantities, while the e_{ij} are random and interest focuses on their variance σ^2.

2.3.5.1 Proportion of Variances

For two normally distributed samples of size n and m, respectively, we estimate the proportion of variances as $\hat{\Pi} = \hat{\sigma}_A^2 / \hat{\sigma}_B^2$. We derive a confidence interval for this estimator by noting that $(n-1)\hat{\sigma}_A^2/\sigma_A^2 \sim \chi_{n-1}^2$ and $(m-1)\hat{\sigma}_B^2/\sigma_B^2 \sim \chi_{m-1}^2$. Then,

$$\frac{\frac{(n-1)\hat{\sigma}_A^2}{\sigma_A^2}/(n-1)}{\frac{(m-1)\hat{\sigma}_B^2}{\sigma_B^2}/(m-1)} = \frac{\hat{\sigma}_A^2/\sigma_A^2}{\hat{\sigma}_B^2/\sigma_B^2} = \frac{\hat{\Pi}}{\Pi} \sim F_{n-1,m-1}$$

has an F-distribution with $n-1$ numerator and $m-1$ denominator degrees of freedom. A $(1-\alpha)$-confidence interval for Π is therefore

$$\left[\frac{\hat{\Pi}}{F_{1-\alpha/2,\,n-1,\,m-1}}, \frac{\hat{\Pi}}{F_{\alpha/2,\,n-1,\,m-1}} \right].$$

For our example, we find the two individual estimates $\hat{\sigma}_A^2 = 3.77$ and $\hat{\sigma}_B^2 = 1.69$ and an estimated proportion $\hat{\Pi} = 2.23$. This yields a 95%-confidence interval for $\hat{\Pi}$ of [0.55, 8.98], which contains the ratio of one (meaning equal variances) and values below and above one. Hence, given the data, we have no evidence that the two true variances are substantially different, even though their estimates differ by a factor of more than 2.

2.3.5.2 Difference in Means of Independent Samples

We estimate the systematic difference Δ between the two kits as

$$\hat{\Delta} = \widehat{\mu_A - \mu_B} = \hat{\mu}_A - \hat{\mu}_B,$$

the difference between the estimates of the respective expected enzyme levels.

For the example, we estimate the two average enzyme levels as $\hat{\mu}_A = 10.32$ and $\hat{\mu}_B = 10.66$ and the difference as $\hat{\Delta} = -0.34$. The estimated difference is not exactly zero, but this might be explainable by measurement error or the natural variation of enzyme levels between mice.

We take account of the uncertainty by calculating the standard error and a 95%-confidence interval for $\hat{\Delta}$. The lower and upper confidence limits then provide information about a potential systematic difference: if the upper limit is below zero, then only negative differences are compatible with the data, and we can conclude that measurements of kit A are systematically lower than measurements of kit B. Conversely, a lower limit above zero indicates kit A yielding systematically higher values than kit B. If zero is contained in the confidence interval, we cannot determine the direction of a potential difference, and it is also plausible that no difference exists.

We already established that the data provides no evidence against the assumption that the two variances are equal. We therefore assume a common variance $\sigma^2 = \sigma_A^2 = \sigma_B^2$, which we estimate by the *pooled variance estimate*

$$\hat{\sigma}^2 = \frac{\hat{\sigma}_A^2 + \hat{\sigma}_B^2}{2} .$$

For our data, $\hat{\sigma}^2 = 2.73$ and the estimated standard deviation is $\hat{\sigma} = 1.65$.

Compared to this standard deviation, the estimated difference is small (about 21%). Whether such a small difference is meaningful in practice depends on the subject matter. A helpful dictum is

A difference which makes no difference is no difference at all. (attributed to William James)

To determine the confidence limits, we first need the standard error of the difference estimate. The two estimates $\hat{\mu}_A$ and $\hat{\mu}_B$ are based on independently selected mice, and are therefore independent. The simple application of the rules for variances then yields

$$\text{Var}(\hat{\Delta}) = \text{Var}(\hat{\mu}_A - \hat{\mu}_B) = \text{Var}(\hat{\mu}_A) + \text{Var}(\hat{\mu}_B) = 2 \cdot \frac{\sigma^2}{n} .$$

The standard error of the difference estimator is therefore

$$\text{se}(\hat{\Delta}) = \sqrt{2 \cdot \frac{\sigma^2}{n}} ,$$

which yields $\widehat{\text{se}}(\hat{\Delta}) = 0.74$. The $(1 - \alpha)$-confidence interval

$$\hat{\Delta} \pm t_{\alpha/2, 2n-2} \cdot \widehat{\text{se}}(\hat{\Delta}) = (\hat{\mu}_A - \hat{\mu}_B) \pm t_{\alpha/2, 2n-2} \cdot \sqrt{2 \cdot \frac{\hat{\sigma}^2}{n}}$$

follows from the previous arguments.

Note that we have $2n$ observations in total, from which we estimate the two means, resulting in $2n - 2$ degrees of freedom for the t-distribution; these are the 18 degrees of freedom of the factor *(Mouse)* in Fig. 2.7.

For our data, we calculate a 95%-confidence interval of $[-1.89, 1.21]$. The confidence interval contains the value zero, and the two kits might indeed produce equivalent values. The interval also contains negative and positive values of substantial magnitude that are possibly relevant in practice. We can therefore only conclude that there is no evidence of a systematic difference between kits, but if such difference

Table 2.4 Enzyme levels for vendors A and B based on two samples from each of 10 mice, one sample per vendor

Mouse	1	2	3	4	5	6	7	8	9	10
A	9.14	9.47	11.14	12.45	10.88	8.49	7.62	13.05	9.67	11.63
B	9.19	9.70	11.12	12.62	11.50	8.99	7.54	13.38	10.94	12.28

A: Experiment structure	**B:** Model parameters

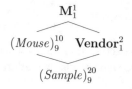

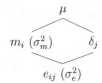

Fig. 2.8 Paired design for estimating average enzyme level difference: 10 mice, each vendor assigned to one of two samples per mouse

existed, it might be large enough to be practically relevant. This non-result should have been avoided with proper experimental design.

2.3.5.3 Difference in Means of Dependent Samples

We find a very different result when we address the same question using the design illustrated in Fig. 1.1C, where we randomly select 10 mice, draw two samples from each, and randomly assign each kit to one sample. The resulting data are given in Table 2.4.

The Hasse diagram for this experiment is shown in Fig. 2.8A. The two factors *(Mouse)* and **Vendor** are now *crossed* and written next to each other. Each sample corresponds to a combination of one mouse and one vendor and is nested in both. Since the mice are randomly selected, their average responses are also random.

The two observations from mouse i are

$$y_{iA} = \mu + \delta_A + m_i + e_{iA} \quad \text{and} \quad y_{iB} = \mu + \delta_B + m_i + e_{iB} \,,$$

where $\text{Var}(m_i) = \sigma_m^2$ and $\text{Var}(e_{ij}) = \sigma_e^2$. The parameters $\delta_A = -\delta_B$ are the treatment effects. The expected response with kit A is then $\mu + \delta_A$, and the variance of any observation is $\text{Var}(y_{ij}) = \sigma_m^2 + \sigma_e^2$. This model corresponds directly to the diagram, as shown in Fig. 2.8B. We estimate the average enzyme level from kit A as $\hat{\mu}_A = \sum_{i=1}^{n} y_{iA}/n$ and from kit B as $\hat{\mu}_B = \sum_{i=1}^{n} y_{iB}/n$. An estimate of their difference $\Delta' = \mu_A - \mu_B$ is then

$$\hat{\Delta}' = \frac{1}{n}\sum_{i=1}^{n}(y_{i,A} - y_{i,B}) = \frac{1}{n}\sum_{i=1}^{n}y_{i,A} - \frac{1}{n}\sum_{i=1}^{n}y_{i,B} = \hat{\mu}_A - \hat{\mu}_B .$$

This is the same estimator as for two independent samples and yields a similar estimated difference of $\hat{\Delta}' = -0.37$ for our example.

The variance of this estimator is very different, however, due to the correlation between each pair of samples,

$$\text{Var}(\hat{\Delta}') = \text{Var}(\hat{\mu}_A) + \text{Var}(\hat{\mu}_B) - 2 \cdot \text{Cov}(\hat{\mu}_A, \hat{\mu}_B) = 2\frac{\sigma_m^2 + \sigma_e^2}{n} - 2\frac{\sigma_m^2}{n} = 2\frac{\sigma_e^2}{n} .$$

In other words, contrasting the two kits within each mouse and then averaging these differences over the 10 mice eliminate the between-mouse variation σ_m^2 from the treatment comparison.

For the data in Table 2.4, we find the two means $\hat{\mu}_A = 10.35$ and $\hat{\mu}_B = 10.73$ together with the variances $\hat{\sigma}_A^2 = 3.1$ and $\hat{\sigma}_B^2 = 3.38$, and a non-zero covariance of $\widehat{\text{Cov}} = 3.16$. The standard error of the difference estimate is then $\widehat{\text{se}}(\hat{\Delta}') = 0.13$, much lower than the previous standard error $\widehat{\text{se}}(\hat{\Delta}) = 0.74$.

The $(1 - \alpha)$-confidence interval for Δ' is

$$\hat{\Delta}' \pm t_{\alpha/2, n-1} \cdot \widehat{\text{se}}(\hat{\Delta}')$$

and is based on $n - 1 = 9$ degrees of freedom (note that we have a single sum with 10 summands, not two independent sums!). The 95%-confidence interval for our data is $[-0.66, -0.08]$ and is much narrower than the interval $[-1.89, 1.21]$ of the previous example due to the substantial increase in precision. It only contains negative values, indicating that kit A consistently yields lower values than kit B. The differences compatible with the data are small, and subject-matter considerations can help determine if such differences are practically relevant. In contrast to the previous experiment, we now arrived at a satisfactory result from which relevant conclusions can be drawn.

Standardized Effect Size for Difference

The difference Δ is an example of a *raw effect size* and has the same units as the original measurements. Subject-matter knowledge often provides information about the relevance of specific effect sizes and might tell us, for example, whether a difference in enzyme levels of more than 0.5 is biologically relevant.

Sometimes the raw effect size is difficult to interpret. This is a particular problem with some current measurement techniques in biology, which give measurements in arbitrary units (a.u.), making it difficult to directly compare results from two experiments. In this case, a unitless *standardized effect size* might be more appropriate. A popular choice is *Cohen's d*, which compares the difference with the standard deviation in the population:

$$d = \frac{\mu_A - \mu_B}{\sigma} \quad \text{estimated by} \quad \hat{d} = \frac{\hat{\mu}_A - \hat{\mu}_B}{\hat{\sigma}} \; .$$

It is a unitless effect size that measures the difference as a multiple of the standard deviation. If $|d| = 1$, then the two means are one standard deviation apart. In the original literature, Cohen suggests that $|d| < 0.2$ should be considered a small effect, $0.2 < |d| < 0.5$ a medium-sized effect, and $0.5 < |d| < 0.8$ a large effect (Cohen 1988), but such definitive categorization should not be taken too literally.

For our example, we calculate $\hat{d} = -0.21$, a difference of 21% of a standard deviation, indicating a small-to-medium effect size. The exact confidence interval for $\hat{d}$ is based on a noncentral t-distribution and cannot be given in closed form (cf. Sect. 2.5). For large enough sample size, we can use a normal confidence interval $\hat{d} \pm z_{\alpha/2} \cdot \widehat{\text{se}}(\hat{d})$ based on an approximation of the standard error (Hedges and Olkin 1985):

$$\widehat{\text{se}}(\hat{d}) = \sqrt{\frac{n_A + n_B}{n_A \cdot n_B} + \frac{\hat{d}^2}{2 \cdot (n_A + n_B)}} \, ,$$

where n_A and n_B are the respective sample sizes for the two groups. For our example, this yields an approximate 95%-confidence interval of $[-1.08, 0.67]$.

2.4 Testing Hypotheses

The underlying question in our kit vendor example is whether the statement $\mu_A = \mu_B$ about the underlying parameters is true or not. We argued that the data would not support this statement if the confidence interval of Δ lies completely above or below zero, excluding $\Delta = 0$ as a plausible difference. *Significance testing* is an equivalent way of using the observations for evaluating the evidence in favor or against a specific *null hypothesis*, such as

$$H_0 : \mu_A = \mu_B \quad \text{or equivalently} \quad H_0 : \Delta = 0 \; .$$

The null hypothesis is a statement about *the true value of one or more parameters*.

2.4.1 The Logic of Falsification

Testing hypotheses follows a logic closely related to that of scientific research in general: we start from a scientific conjecture or theory to explain a phenomenon. From the theory, we derive observable consequences (or predictions) that we test in an experiment. If the predicted consequences are in substantial disagreement with those experimentally observed, then the theory is deemed implausible and is *rejected* or *falsified*. If, on the other hand, prediction and observation are in agreement, then the experiment fails to reject the theory. However, this does not mean that the theory is proven or verified since the agreement might be due to chance, the experiment not specific enough, or the data too noisy to provide sufficient evidence against the theory:

Absence of evidence is not evidence of absence.

It is instructive to explicitly write down this logic more formally. The correctness of the conjecture C implies the correctness of the prediction P:

$$C \text{ true} \implies P \text{ true} .$$

We say C is *sufficient* for P. We now perform an experiment to determine if P is true. If the experiment supports P, we do not gain much, since we cannot invert the implication and conclude that the conjecture C must also be true. This is because many other possible conditions unrelated to our conjecture might also lead to the predicted outcome.

However, if the experiment shows the prediction P to be false, then we can invoke the *modus tollens*

$$P \text{ false} \implies C \text{ false}$$

and conclude that C cannot be true. We say P is *necessary* for C. There is thus an asymmetry in the relation of 'P is true' and 'P is false' toward the correctness of C, and we can falsify a theory (at least in principle), but never fully verify it. The philosopher Karl Popper argued that falsification is a cornerstone of science, and falsifiability (in principle) of conjectures separates a scientific theory from a non-scientific one (Popper 1959).

Statistical testing of hypotheses follows a similar logic, but adds a probabilistic argument to quantify the (dis-)agreement between hypothesis and observation. The data may provide evidence to reject the null hypothesis, but can never provide evidence for accepting the null hypothesis. We therefore formulate a null hypothesis for the 'undesired' outcome such that if we don't reject it, nothing is gained and we don't have any clue from the data how to proceed in our investigation. If the data provides evidence to reject the hypothesis, however, we can reasonably exclude it as a possible explanation of the observed data.

To appraise the evidence that our data provides against the null hypothesis, we need to take the random variation in the data into account. Instead of a yes/no answer, we can then only argue that "if the hypothesis is true, then it is (un-)likely that data like ours are observed." This is precisely the argument from the hypothesis to the observable consequences, but with a probabilistic twist.

2.4.2 The t-Test

For our example, we know that if the true difference Δ is zero, then the estimated difference $\hat{\Delta}$ divided by its standard error has a t-distribution. This motivates the well-known t-*statistic*

$$T = \frac{\hat{\Delta}}{\widehat{se}(\hat{\Delta})} = \frac{\hat{\Delta}}{\sqrt{2 \cdot \hat{\sigma}^2/n}} \sim t_{2n-2} . \tag{2.1}$$

Thus, our conjecture is "the null hypothesis is true and $\Delta = 0$" from which we derive the prediction "the observed test statistic T based on the two sample averages follows a t-distribution with $2n - 2$ degrees of freedom".

For our data, we compute a t-statistic of $t = -0.46$, based on the two means 10.32 for vendor A, 10.66 for vendor B, their difference $\hat{\Delta} = -0.34$, and the resulting standard error of 0.74 of the difference on 18 degrees of freedom. The estimated difference in means is expressed by t as about 46% of a standard error, and the sign of t indicates that measurements with kit A might be lower than those with kit B.

The test statistic remains the same for the case of paired data as in Fig. 1.1C, with standard error calculated as in Sect. 2.3.5.3; this is known as a *paired t-test*.

2.4.3 p-Values and Statistical Significance

p-values

If we assume that the hypothesis H_0 is true, we can compute the probability that our test statistic T exceeds any given value in either direction using its known distribution. Calculating this probability for the observed value t provides us with a quantitative measure of the evidence that the data provide against the hypothesis. This probability is called the *p-value*

$$p = \mathbb{P}(|T| \geq |t| \,|\, H_0 \text{ true}) .$$

Because our test statistic is random (it is a function of the random samples), the p-value is also a random variable. If the null hypothesis is true, then p has a uniform distribution between 0 and 1.

For our example, we compute a p-value of 0.65, and we expect a t-statistic that deviates from zero by 0.46 or more in either direction in 65 out of 100 cases whenever the null hypothesis is true. We conclude that based on the variation in the data and the sample size, observing a difference of at least this magnitude is very likely and there is no evidence against the null hypothesis.

A small p-value is considered indicative of H_0 being false, since it is unlikely that the observed (or larger) value of the test statistic would occur if the hypothesis were true. This leads to a dichotomy of explanations for small p-values: either the data led to the large value just by chance, or the null hypothesis is wrong and the test statistic does not follow the predicted distribution. On the other hand, a large p-value might occur either because H_0 is true or because H_0 is false but our data did not yield sufficient information to detect the true difference in means.

We may still decide to ignore a large p-value and nevertheless move ahead and assume that H_0 is wrong; but the argument for doing so cannot rest on the single experimental result and its statistical analysis, but must include external arguments such as subject-matter considerations (e.g., about plausible effects) or outcomes of related experiments.

If possible, we should always combine an argument based on a p-value with the observed effect size: a small p-value (indicating that the effect found can be distinguished from noise) is only meaningful if the observed effect size has a practically relevant magnitude. Even a large effect size might yield a large p-value if the sample size is small and the variation in the data is large. In our case, the estimated difference of $\hat{\Delta} = -0.34$ is small compared to the standard error $\widehat{se}(\hat{\Delta}) = 0.74$ and cannot be distinguished from random fluctuation. That our test does not provide evidence for a difference, however, does not mean it does not exist.

Conversely, a low p-value means that we were able to distinguish the observed difference from a random fluctuation, but the detected difference might have a tiny and irrelevant effect size if the random variation is low or the sample size large. However, this provides only one (crucial) piece of the argument why we suggest a particular interpretation and rule out others. It does not excuse us from proposing a reasonable interpretation of the *whole* investigation and other experimental data.

Statistical Significance

We sometimes use a pre-defined *significance level* α against which we compare the p-value. If $p < \alpha$, then we consider the p-value low enough to *reject* the null hypothesis and we call the test *statistically significant*; the hypothesis is *not rejected* otherwise, a result called *not significant*. The level α codifies our judgement of what constitutes a 'just by chance' event, and is the probability to incorrectly reject a true null hypothesis. There is no theoretical justification for any choice of this threshold, but for as much historical as practical reasons, values of $\alpha = 5\%$ or $\alpha = 1\%$ are commonly used. In our example, $p = 0.65$ far exceeds both thresholds and the test outcome is thus deemed not significant.

All cautionary notes for p-values apply to significance statements as well. In particular, just because a test found a significant difference does not imply that the difference is practically meaningful:

Statistical significance does not imply scientific significance.

Moreover, the p-value or a statement of significance both concern a single experiment, but a single significant result is not enough to establish the scientific significance of a result:

> *In order to assert that a natural phenomenon is experimentally demonstrable we need, not an isolated record, but a reliable method of procedure. In relation to the test of significance, we may say that a phenomenon is experimentally demonstrable when we know how to conduct an experiment which will rarely fail to give us a statistically significant result.*
> (Fisher 1971)

as R. A. Fisher rightly insisted.

Instead of comparing the p-value with the significance level, we can equivalently compare the observed test statistic t with the corresponding quantiles of the null-distribution. We then reject the null hypothesis if $T < t_{\alpha/2, 2n-2}$ (kit A lower) or $T > t_{1-\alpha/2, 2n-2}$ (kit B lower). For our example with $\alpha = 5\%$, the thresholds are $t_{0.025, 18} = -2.1$ and $t_{0.975, 18} = 2.1$. Our observed t-statistic $t = -0.46$ lies within these thresholds and the hypothesis is not rejected.

Error Probabilities

Any statistical test has four possible outcomes under the *significant/non-significant* dichotomy, which we summarize in Table 2.5. A significant result leading to rejection of H_0 is called a *positive*. If the null hypothesis is indeed false, then it is a *true positive*, but if the null hypothesis is indeed true, it was incorrectly rejected: a *false positive* or *type-I error*. The significance level α is the probability of a false positive and by choosing a specific value, we can control the type-I error of our test. The *specificity* of the test is the probability $1 - \alpha$ that we correctly do not reject a true null hypothesis.

Conversely, a non-significant result is called a *negative*. Not rejecting a true null hypothesis is a *true negative*, and incorrectly not rejecting a false null hypothesis is a *false negative* or a *type-II error*. The probability of a false negative test outcome is denoted by β. The *power* or *sensitivity* of the test is the probability $1 - \beta$ that we correctly reject the null hypothesis. The larger the true effect size, the more power a test has to detect it. *Power analysis* allows us to determine the test parameters (most importantly, the sample size) to provide sufficient power for detecting an effect size deemed scientifically relevant.

Everything else being equal, the two error probabilities α and β are adversaries: by lowering the significance level α, larger effects are required to reject the null hypothesis, which simultaneously increases the probability β of a false negative.

This has an important consequence: if our experiment has low power (it is *under-powered*), then only large effect sizes yield statistical significance. A significant

Table 2.5 The four possible outcomes of a hypothesis test

	H_0 true	H_0 false
No rejection	*true negative* (TN) *specificity* $1 - \alpha$	*false negative* (FN) β
Rejection	*false positive* (FP) α	*true positive* (TP) *sensitivity* or *power* $1 - \beta$

p-value and a large effect size then tempt us to conclude that we reliably detected a substantial difference, even though we must expect such a result with probability α in the case that *no* difference exists.

Under the repeated application of a test with new data, α and β are the expected frequencies—the *false positive rate*, respectively, the *false negative rate*—of the two types of error. This interpretation is helpful in scenarios such as quality control. On the other hand, we usually conduct only a limited number of repeated tests in scientific experimentation. Here, we can interpret α and β as quantifying the *severity* of a statistical test—its hypothetical capability to detect the desired difference in a planned experiment. This is helpful for planning experiments to ensure that the data generated will likely provide sufficient evidence against an incorrect scientific hypothesis.

Rejection Regions and Confidence Intervals

Instead of rejecting H_0 if the test statistic T exceeds the respective t-quantiles, we can also determine the corresponding limits on the estimated difference $\hat{\Delta}$ directly and reject if

$$\hat{\Delta} < t_{\alpha/2,\, 2n-2} \cdot \sqrt{2} \cdot \hat{\sigma}/\sqrt{n} \quad \text{or} \quad \hat{\Delta} > t_{1-\alpha/2,\, 2n-2} \cdot \sqrt{2} \cdot \hat{\sigma}/\sqrt{n}\,.$$

The corresponding set of values is called the *rejection region* $\mathcal{R}$ and we reject H_0 whenever the observed difference $\hat{\Delta}$ falls inside $\mathcal{R}$. In our example, $\mathcal{R}$ consists of the two sets $(-\infty, -1.55)$ and $(1.55, +\infty)$. Since $\hat{\Delta} = -0.34$ does not fall into this region, we do not reject the null hypothesis.

There is a one-to-one correspondence between the rejection region $\mathcal{R}$ for a significance level α, and the $(1 - \alpha)$-confidence interval of the estimated difference. The estimated effect $\hat{\Delta}$ lies inside the rejection region if and only if the value 0 is outside its $(1 - \alpha)$-confidence interval. This equivalence is illustrated in Fig. 2.9. In

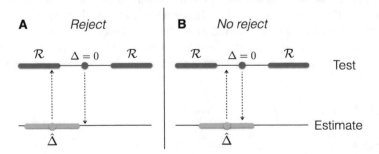

Fig. 2.9 Equivalence of testing and estimation. **A** Estimate inside rejection region and null value outside the confidence interval. **B** Estimate outside rejection region, confidence interval contains null value. Top: null value (dark gray dots) and rejection region (dark gray shades); bottom: estimate $\hat{\Delta}$ (light gray dots) and confidence interval (light gray shades)

hypothesis testing, we put ourselves on the value under H_0 (e.g., $\Delta = 0$) and check if the estimated value $\hat{\Delta}$ is so far away that it falls inside the rejection region. In estimation, we put ourselves at the estimated value $\hat{\Delta}$ and check if the data provide no objection against the hypothesized value under H_0, which then falls inside the confidence interval. Our previous argument that the 95%-confidence interval of $\hat{\Delta}$ does contain zero is therefore equivalent to our current result that $H_0 : \Delta = 0$ cannot be rejected at the 5% significance level.

2.4.4 Four Additional Test Statistics

We briefly present four more significance test statistics for testing expectations and variances. Their distributions under the null hypothesis are all based on normally distributed data.

z-Test of Mean Difference

For large sample sizes n, the t-distribution quickly approaches a standard normal distribution. If $2n - 2$ is 'large enough' (typically, $n > 20$ suffices), we can then compare the test statistic

$$T = \frac{\hat{\Delta}}{\widehat{\text{se}}(\hat{\Delta})}$$

with the quantiles z_α of the standard normal distribution instead of the exact t-quantiles, and reject H_0 if $T > z_{1-\alpha/2}$ or $T < z_{\alpha/2}$. This test is sometimes called a z-test. There is usually no reason to forego the exactness of the t-test, however.

For our example, we reject the null hypothesis at a 5% level if the test statistic T is below the approximate $z_{0.025} = -1.96$, respectively, and above $z_{0.975} = 1.96$, as compared to the exact thresholds $t_{0.025,18} = -2.1$ and $t_{0.975,18} = 2.1$.

χ^2-Test of Variance

Given an estimate $\hat{\sigma}^2$ from normally distributed observations based on d degrees of freedom, we can use a χ^2-test to test $H_0 : \sigma^2 = \sigma_0^2$, the null hypothesis that the variance is equal to a specific value σ_0^2. The test statistic

$$T = d \cdot \frac{\hat{\sigma}^2}{\sigma_0^2}$$

has a χ^2-distribution with d degrees of freedom if the null hypothesis is true. We reject H_0 if $T < \chi^2_{\alpha/2,d}$ or $T > \chi^2_{1-\alpha/2,d}$. It is sometimes sufficient to test if the true variance is smaller than σ_0^2 such that we establish an upper bound for the variance. We reject this hypothesis if $T > \chi_{1-\alpha,d}^2$ (note that we use α here and not $\alpha/2$), but the true variance can be arbitrarily smaller than σ_0^2 if we do not reject the null hypothesis.

For our example data for vendor A, we have $n = 10$ normally distributed samples and the variance estimate is based on $n - 1 = 9$ degrees of freedom. We reject the hypothesis $H_0 : \sigma_A^2 = 3$ at a 5% level if $T = 9\hat{\sigma}_A^2/3 < 2.7$ or $T > 19.02$. We calculate $t = 11.31$ and find a p-value of 0.75.

F-Test of Variance Ratio

If we have two variance estimates $\hat{\sigma}_A^2$ and $\hat{\sigma}_B^2$, calculated from normally distributed samples and based on d_A and d_B degrees of freedom, respectively, we can test their equality using the test statistic

$$T = \frac{\hat{\sigma}_A^2}{\hat{\sigma}_B^2} .$$

Under the null hypothesis $H_0 : \sigma_A^2 = \sigma_B^2$, which is equivalent to $H_0 : \sigma_A^2/\sigma_B^2 = 1$, the test statistic has an F-distribution with d_A numerator and d_B denominator degrees of freedom. We reject H_0 if $T < F_{\alpha/2,d_A,d_B}$ or $T > F_{1-\alpha/2,d_A,d_B}$.

In our example, we have $\hat{\sigma}_A^2 = 3.77$ and $\hat{\sigma}_B^2 = 1.69$ based on $d_A = d_B = n - 1 = 9$ degrees of freedom each. This leads to an observed test statistic of $t = 2.23$. The critical values at a 5% level are $F_{0.025,9,9} = 0.25$ and $F_{0.975,9,9} = 4.03$ and the resulting p-value is 0.25.

Levene's Test for Equality of Multiple Variances

In subsequent chapters, we frequently make the assumption that the variances of several groups of data are all equal; this is called *variance homogeneity*. For k groups, this corresponds to the null hypothesis

$$H_0 : \sigma_1^2 = \cdots = \sigma_k^2 .$$

This hypothesis can be tested using *Levene's test*.

With observations $y_{ij}, i = 1 \ldots k$, and $j = 1 \ldots n_i$, we define the deviations from each group mean $\bar{y}_i$ as $r_{ij} = |y_{ij} - \bar{y}_i|$; these correspond to absolute residuals. Let $\bar{r}_i = \sum_{j=1}^{n_i} r_{ij}$ and $\bar{r} = \sum_{i=1}^{k} \sum_{j=1}^{n_i} r_{ij}$ be the average residuals in group i, respectively overall, and denote by $N = n_1 + \cdots n_k$ the overall sample size.

The test statistic

$$W = \frac{\sum_{i=1}^{k} n_i \cdot (\bar{r}_i - \bar{r})/(k - 1)}{\sum_{i=1}^{k} \sum_{j=1}^{n_i} (r_{ij} - \bar{r}_i)/(N - k)}$$

compares the variance of the average residuals between groups to the variance of the residuals within the groups, and has an F-distribution with $k - 1$ numerator and $N - k$ denominator degrees of freedom if H_0 is true.

2.5 Notes and Summary

Notes

Most material in this chapter is covered by any introductory book on statistics. Two historical examples are Fisher's *Statistical Methods for Research Workers* (Fisher 1925) and Snedecor and Cochran's *Statistical Methods* (Snedecor and Cochran 1989), and a very readable account of the historical development of statistics is Salsburg (2002). A recent introductory text is Wolfe and Schneider (2017), and a very broad treatment of many methods is Wasserman (2014). Other texts emphasize using R (Dalgaard 2008; Field et al. 2012; Shahbaba 2012). We only covered a statistical approach often called *frequentist*; an alternative approach is *Bayesian statistics*, covered at length in Gelman (2013).

There has been a re-emphasis on estimates and their uncertainties rather than testing in recent years; a standard account of this approach is Altman et al. (2000b). In particular, there is a strong incentive to abandon the *significant/non-significant* dichotomy in hypothesis testing in favor of *p*-values, estimates, and confidence intervals; in 2019, the American Statistical Association devoted a special issue entitled *Statistical Inference in the 21st Century: A World Beyond p < 0.05* to this topic (Wasserstein et al. 2019). The relation of sample size, confidence intervals, and power is discussed in Greenland et al. (2016). An insightful discussion of statistical significance is Cox (2020). An interesting perspective on statistical inference based on the notion of *severity* is given in Mayo (2018).

Standardized effect sizes are popular in fields like psychology and social sciences, but have also been advocated for biological research (Nakagawa and Cuthill 2007). The bootstrap method was proposed in Efron (1979) and DiCiccio and Efron (1996) is a more recent review of confidence interval estimation.

The diagrams in Figs. 2.4, 2.7, and 2.8 are examples of *Hasse diagrams*. Their use to describe statistical models and analyses began with Tjur (1984), and variants have been proposed regularly for visualizing experimental designs and automating their analysis (Taylor and Hilton 1981; Brien 1983; Bergerud 1996; Darius et al. 1998; Vilizzi 2005; Großmann 2014; Bate and Chatfield 2016, b). A review of recent developments is Bailey (2020). Their use for planning and discussing experimental designs in non-technical terms is emphasized by Lohr (1995), but only two previous textbooks make use of them (Oehlert 2000; Bailey 2008).

Exact confidence intervals for Cohen's d

We can calculate an exact confidence interval for an estimated standardized effect size $\hat{d}$ using the noncentral t-distribution with estimated noncentrality parameter $\hat{\eta} = \hat{d} \cdot \sqrt{n/2}$. We calculate a $(1 - \alpha)$-confidence interval for $\hat{\eta}$ and then transform it into a corresponding interval for $\hat{d}$ by multiplying the lower and upper confidence limits by $\sqrt{2/n}$. To find the confidence interval for $\hat{\eta}$, we need to find the noncentrality parameters η_{lcl} and η_{ucl} such that

$$\mathbb{P}(T_{ucl} \leq \hat{\eta}) = 1 - \alpha/2 \text{ for } T_{ucl} \sim t_{2n-2}(\eta_{ucl})$$
$$\mathbb{P}(T_{lcl} \leq \hat{\eta}) = \alpha/2 \text{ for } T_{lcl} \sim t_{2n-2}(\eta_{lcl}) \ .$$

The following R code implements this strategy:

```
# Calculate difference between estimated and given ncp
# eta: noncentrality parameter for t-distribution
# etahat: ncp estimate for which CI is computed
# df: degrees of freedom for t-distribution
# level: confidence level
nt = function(eta, etahat, df, level) {
    dist = ( pt(q=etahat, df=df, ncp=eta) - level )^2
    return(dist)
}
etahat = dhat * sqrt(n/2); df = 2*n-2; alpha=0.05
# Find 2.5% and 97.5% quantiles for eta (p is starting value)
# by minimization of nt() over ncp parameter
# nlm(): finds minimum of nt() over parameter 'ncp'
ucl.eta = nlm(nt,p=etahat,etahat=etahat,df=df,level=alpha/2)
lcl.eta = nlm(nt,p=etahat,etahat=etahat,df=df,level=1-alpha/2)
# Transform CI for eta into CI for d
ucl.d = ucl.eta$estimate * sqrt(2/n)
lcl.d = lcl.eta$estimate * sqrt(2/n)
```

Using R

Basic R covers most frequently used distributions; for a distribution X, the functions dX() and pX() provide the density and distribution function, qX() calculates quantiles, and rX() provides a random number generator. For example, the χ^2-distribution has functions pchisq(), dchisq(), qchisq(), and rchisq(). Noncentral distributions are accessed using the ncp= argument in the corresponding functions, e.g., qt(p=0.05, df=6, ncp=1) gives the 5% quantile of a noncentral t-distribution with $\eta = 1$ and six degrees of freedom.

Data is most conveniently brought into a *data-frame*, which is a rectangular table whose columns can be accessed by name. Our vendor example uses a data-frame with columns y for the enzyme level (a number per row), and vendor to encode the vendor (with entry either A or B). The *tidyverse* encompasses a number of packages to comfortably work with data-frames, for example, dplyr and tidyr. A good introduction to this framework is Wickham and Grolemund (2016a), which is also freely available online.

Standard estimators are mean() for the expectation, and var() and sd() for variance and standard deviation. Covariances and correlations are calculated by cov() and cor(). Confidence intervals are not readily available and are computed manually; for means and mean differences, we can use a linear model of the form y~1, respectively, y~d (d encoding the two groups) in `lm()`, and apply `confint()` to the resulting model.

The t-test is computed using `t.test()`; the option `paired=TRUE` calculates the paired t-test, and `var.equal=TRUE` enforces that variances in both samples are considered equal. The χ^2-test and the F-test are available as `chisq.test()`, respectively, `var.test()`, and Levene's test as `leveneTest()` from package `car`.

The `effectsize` package provides `cohens_d()` for calculating Cohen's d with an exact confidence interval.

Summary

Probability distributions describe random outcomes, such as measurements with noise; they are characterized by the cumulative distribution and density functions. An α-quantile of a distribution is the value such that a proportion of α of possible realizations will fall below the quantile; this plays a prominent role in constructing confidence intervals and rejection regions.

An estimator takes a random sample and calculates the best guess for the desired parameter. It has a mean and standard deviation (called the standard error) and precision and accuracy describe important properties. The precision of an estimate is given by the standard error, and is used to construct the confidence interval that contains all values for the parameter that are compatible with the data.

Statements about parameters can be made in the form of statistical hypotheses, which are then tested against available data. The hypothesis is rejected (or falsified) if the deviation between the expected distribution and the observed outcome of a test statistic is larger than a given threshold, defined by the significance level. The lowest significance level not resulting in rejection is the p-value, the probability to see the observed deviation or larger if the hypothesis is in fact true.

For scientific inference, hypothesis tests are usually of secondary interest, and estimates of relevant parameters and effect sizes should be given instead, together with their confidence intervals. The hypothesis testing framework is useful, however, in the planning stage of an experiment to determine sample sizes, among others.

References

Altman, D. G. et al. (2000). Statistics with Confidence. 2nd. John Wiley & Sons, Inc.

Bailey, R. A. (2008). Design of Comparative Experiments. Cambridge University Press.

Bailey, R. A. (2020). "Hasse diagrams as a visual aid for linear models and analysis of variance". In: Communications in Statistics - Theory and Methods, pp. 1–34.

Bate, S. T. and M. J. Chatfield (2016a). "Identifying the Structure of the Experimental Design". In: Journal of Quality Technology 48.4, pp. 343–364.

Bate, S. T. and M. J. Chatfield (2016b). "Using the Structure of the Experimental Design and the Randomization to Construct a Mixed Model". In: Journal of Quality Technology 48.4, pp. 365–387.

Bergerud, W. A. (1996). "Displaying factor relationships in experiments". In: The American Statistician 50.3, pp. 228–233.

Brien, C. J. (1983). "Analysis of Variance Tables Based on Experimental Structure". In: Biometrics 39.1, pp. 53–59.

Cohen, J. (1988). Statistical Power Analysis for the Behavioral Sciences. 2nd. Lawrence Erlbaum Associates, Hillsdale.

Cox, D. R. (2020). "Statistical significance". In: Annual Review of Statistics and Its Application 7, pp. 1.1–1.10.

Dalgaard, P. (2008). Introductory Statistics with R. Statistics and Computing. Springer New York.

Darius, P. L., W. J. Coucke, and K. M. Portier (1998). "A Visual Environment for Designing Experiments". In: Compstat, pp. 257–262.

DiCiccio, T. J. and B. Efron (1996). "Bootstrap confidence intervals". In: Statistical Science 11.3, pp. 189–228.

Efron, B. (1979). "Bootstrap Methods: another look at the jackknife". In: Annals of Statistics 7.11, pp. 1–26.

Field, A., J. Miles, and Z. Field (2012). Discovering Statistics Using R. SAGE Publications Ltd.

Fisher, R. A. (1925). Statistical Methods for Research Workers. 1st. Oliver & Boyd, Edinburgh.

Fisher, R. A. (1971). The Design of Experiments. 8th. Hafner Publishing Company, New York.

Gelman, A. et al. (2013). Bayesian Data Analysis. 3rd. Taylor & Francis.

Greenland, S. et al. (2016). "Statistical tests, P values, confidence intervals, and power: a guide to misinterpretations". In: European Journal of Epidemiology 31.4, pp. 337–350.

Großmann, H. (2014). "Automating the analysis of variance of orthogonal designs". In: Computational Statistics and Data Analysis 70, pp. 1–18.

Hedges, L. and I. Olkin (1985). Statistical methods for meta-analysis. Academic Press.

Lohr, S. L. (1995). "Hasse diagrams in statistical consulting and teaching". In: The American Statistician 49.4, pp. 376–381.

Mayo, D. G. (2018). Statistical Inference as Severe Testing. Cambridge University Press.

Nakagawa, S. and I. C. Cuthill (2007). "Effect size, confidence interval and statistical significance: a practical guide for biologists." In: Biological Reviews of the Cambridge Philosophical Society 82.4, pp. 591–605.

Oehlert, G. W. (2000). A First Course in Design and Analysis of Experiments. W. H. Freeman.

Popper, K. R. (1959). The Logic of Scientific Discovery. Routledge.

Salsburg, D. (2002). The Lady Tasting Tea. Holt Paperbacks.

Shahbaba, B. (2012). Biostatistics with R. Springer New York.

Snedecor, G. W. and W. G. Cochran (1989). Statistical Methods. 8th. Iowa State University Press.

Taylor, W. H. and H. G. Hilton (1981). "A Structure Diagram Symbolization for Analysis of Variance". In: The American Statistician 35.2, pp. 85–93.

Tjur, T. (1984). "Analysis of variance models in orthogonal designs". In: International Statistical Review 52.1, pp. 33–65.

Vilizzi, L. (2005). "The linear model diagram: A graphical method for the display of factor relationships in experimental design". In: Ecological Modelling 184.2-4, pp. 263–275.

Wasserman, L. (2004). All of Statistics. Springer Texts in Statistics. Springer New York.

Wasserstein, R. L., A. L. Schirm, and N. A. Lazar (2019). "Moving to a World Beyond "p < 0.05"". In: The American Statistician 73.sup1, pp. 1–19.

Wickham, H. and G. Grolemund (2016). R for Data Science. O'Reilly.

Wolfe, D. A. and G. Schneider (2017). Intuitive Introductory Statistics. Springer.

Chapter 3
Planning for Precision and Power

3.1 Introduction

In our original two-vendor example, we found that 20 mice were insufficient to conclude if the two kits provide comparable measurements on average or not. We now discuss several methods for increasing precision and power by reducing the standard error: (i) balancing allocation of a fixed number of samples to the treatment groups; (ii) reducing the standard deviation of the responses; and (iii) increasing the sample size. The sample size required for desired precision and power is determined by *power analysis*.

3.2 Balancing Allocation

Without comment, we always used a *balanced allocation*, where the same number of experimental units is allocated to each treatment group. This choice seems intuitively sensible, and we quickly confirm that it indeed yields the highest precision and power in our example. We will later see that unbalanced allocation not only decreases precision, but might prevent the estimation of relevant treatment effects altogether in more complex designs.

We denote by n_A and n_B the number of mice allocated to kits A and B, respectively. The standard error of our estimate is then

$$\mathrm{se}(\hat{\mu}_A - \hat{\mu}_B) = \sqrt{\frac{1}{n_A} + \frac{1}{n_B}} \cdot \sigma \, ,$$

where we estimate the two expectations by $\hat{\mu}_A = \sum_{i=1}^{n_A} y_{i,A}/n_A$ and correspondingly for $\hat{\mu}_B$.

© Springer Nature Switzerland AG 2021
H.-M. Kaltenbach, *Statistical Design and Analysis of Biological Experiments*,
Statistics for Biology and Health, https://doi.org/10.1007/978-3-030-69641-2_3

For fixed total sample size $n_A + n_B$, this standard error is minimal for a balanced allocation with treatment groups of equal size $n_A = n_B$, provided the variance σ^2 is identical in both treatment groups. The more unbalanced the allocation is, the larger the standard error will become.

To illustrate, we consider two experimental designs with a total sample size of $n_A + n_B = 20$: first, we assign a single mouse to vendor B ($n_B = 1$), and the remaining mice to vendor A ($n_A = 19$). Then,

$$
\mathrm{se}_{19,1} = \sqrt{\frac{1}{19} + \frac{1}{1}} \cdot \sigma = 1.026\,\sigma
$$

and the standard error is even higher than the dispersion in the population! However, if we assign the mice equally ($n_A = n_B = 10$), we get a substantially lower standard error of

$$
\mathrm{se}_{10,10} = \sqrt{\frac{1}{10} + \frac{1}{10}} \cdot \sigma = 0.45\,\sigma \ .
$$

The *relative efficiency* of the two designs

$$
\mathrm{RE} = \frac{\mathrm{se}_{19,1}^2}{\mathrm{se}_{10,10}^2} = \left(\frac{1.026\,\sigma}{0.45\,\sigma}\right)^2 \approx 5.2
$$

allows us to directly compare the precision of the two allocation strategies. It is the increase in sample size needed for the first experiment to match the precision of the second. Here, the same unbalanced allocation would require about five times more mice to match the precision of the balanced design. This would mean using at least 5 mice for vendor B and 95 mice for vendor A (100 mice in total). Dividing the experimental material inaptly results in a substantial loss of precision, which is very costly to make up for.

If the two treatment groups have very different standard deviations σ_A and σ_B, then the standard error is

$$
\mathrm{se}(\hat{\mu}_A - \hat{\mu}_B) = \sqrt{\frac{\sigma_A^2}{n_A} + \frac{\sigma_B^2}{n_B}} \ .
$$

The standard error is minimal if $\sigma_A/n_A \approx \sigma_B/n_B$ and we should allocate the samples proportionally to the standard deviation in each group. For $n_A + n_B = 20$ mice and $\sigma_A = 2.8$ twice as large as $\sigma_B = 1.4$, we then allocate $n_A = 13$ and $n_B = 7$ mice, respectively, and use $n_A = 16$ and $n_B = 4$ mice if the standard deviation in group A is $\sigma_A = 5.6$. Very disparate standard errors in the different treatment groups are often a sign that the treatment groups are different in ways other than the assigned treatment alone; such a situation will require more care in the statistical analysis.

3.3 Reducing the Standard Error

We can increase precision and power by reducing the standard deviation σ of our response values. This option is very attractive, since reducing σ to one-half will also cut the standard error to one-half and increase the value of the t-statistic by two, without altering the necessary sample size.

Recall that the standard deviation describes how dispersed the measured enzyme levels of individual mice are around the population mean in each treatment group. This dispersion contains the biological variation σ_m from mouse to mouse, and the variation σ_e due to within-mouse variability and measurement error, such that

$$\sigma^2 = \sigma_m^2 + \sigma_e^2 \quad \text{and} \quad \text{se}(\hat{\Delta}) = \sqrt{2 \cdot \left(\frac{\sigma_m^2 + \sigma_e^2}{n} \right)}.$$

3.3.1 Sub-sampling

If the variance σ^2 is dominated by variation between samples of the same mouse, we can reduce the standard error by taking multiple samples from each mouse. Averaging their measured enzyme levels estimates each mouse's response more precisely. Since the number of experimental units does not change, the measurements from the samples are sometimes called *technical replicates* as opposed to *biological replicates*. If we take m samples per mouse, this reduces the variance of these new responses to $\sigma_m^2 + \sigma_e^2/m$, and the standard error to

$$\text{se}(\hat{\Delta}) = \sqrt{2 \cdot \left(\frac{\sigma_m^2}{n} + \frac{\sigma_e^2}{m \cdot n} \right)}.$$

We can employ the same strategy if the measurement error is large, and we decrease its influence on the standard error by taking r measurements of each of m samples.

This strategy is called *sub-sampling* and is only successful in increasing precision and power substantially if σ_e is not small compared to the between-mouse variation σ_m, since the contribution of σ_m on the standard error only depends on the number n of mice, and not on the number m of samples per mouse. In biological experiments, the biological (mouse-to-mouse) variation is typically much larger than the technical (sample-to-sample) variation and sub-sampling is of very limited use. Indeed, a very common mistake is to ignore the difference between technical and biological replicates and treat all measurements as biological replicates. This flaw is known as *pseudo-replication* and leads to overestimating the precision of an estimate and thus to much shorter, incorrect, confidence intervals, and to overestimating the power of a test, with too low p-values and high probability of false positives (Hurlbert 1984, 2009).

For our examples, the between-mouse variance is $\sigma_m^2 = 1.9$ and much larger than the within-mouse variance $\sigma_e^2 = 0.1$. For $n = 10$ mice per treatment group and $m = 1$ samples per mouse, the standard error is 0.89. Increasing the number of samples to $m = 2$ reduces this error to 0.88 and further increasing to an unrealistic $m = 100$ only reduces the error down to 0.87. In contrast, using 11 instead of 10 mice per treatment reduces the standard error already to 0.85.

3.3.2 Narrowing the Experimental Conditions

We can reduce the standard deviation σ of the observations by tightly controlling experimental conditions, for example, by keeping temperatures and other environmental factors at a specific level, reducing the diversity of the experimental material, and similar measures.

For our examples, we can reduce the between-mouse variation σ_m by restricting attention to a narrower population of mice. If we sample only female mice within a narrow age span from a particular laboratory strand, the variation might be substantially lower than for a broader selection of mice.

However, by ever more tightly controlling the experimental conditions, we simultaneously restrict the generalizability of the findings to the narrower conditions of the experiment. Claiming that the results should also hold for a wider population (e.g., other age cohorts, male mice, or more than one laboratory strand) requires external arguments and cannot be supported by the data from our experiment.

3.3.3 Blocking

Looking at our experimental question more carefully, we discovered a simple yet highly efficient technique to completely remove the largest contributor σ_m from the standard error. The key observation is that we apply each kit to a sample from a mouse and not to the mouse directly. Rather than taking one sample per mouse and randomly allocating it to either kit, we can also take two samples from each mouse, and randomly assign kit A to one sample, and kit B to the other sample (Fig. 1.1C). For each mouse, we estimate the difference between vendors by subtracting the two measurements, and average these differences over the mice. Each individual difference is measured under very tight experimental conditions (within a single mouse); provided the differences vary randomly and independently of the mouse (note that the measurements themselves of course depend on the mouse!), their average yields a *within-mouse* estimate for $\hat{\Delta}$.

Such an experimental design is called a *blocked design*, where the mice are blocks for the samples: they group the samples into pairs belonging to the same mouse, and treatments are independently randomized to the experimental units within each group. This effectively removes the variation between blocks from the treatment comparison, as we saw in Sect. 2.3.5.3.

If we consider our paired-vendor example again, each observation has variance $\sigma_m^2 + \sigma_e^2$, and the two treatment group mean estimates $\hat{\mu}_A$ and $\hat{\mu}_B$ both have variance $(\sigma_m^2 + \sigma_e^2)/n$. However, the estimate $\hat{\Delta} = \hat{\mu}_A - \hat{\mu}_B$ of their difference only has variance $2 \cdot \sigma_e^2/n$, and the between-mouse variance σ_m^2 is completely eliminated from this estimate. This is because each observation from the same mouse is equally affected by any systematic deviation that exists between the specific mouse and the overall average, and this deviation therefore cancels if we look at differences between observations from the same mouse.

For $\sigma_m^2 = 1.9$ and $\sigma_e^2 = 0.1$, for example, blocking by mouse reduces the expected standard error from 0.89 in the original two-vendor experiment to 0.2 in the paired-vendor experiment. Simultaneously, the experiment size is reduced from 20 to 10 mice, while the number of observations remains the same. Importantly, the samples—and not the mice—are the experimental units in this experiment, since we randomly assign kits to samples and not to mice. In other words, we still have 20 experimental units, the same as in the original experiment.

The relative efficiency between unblocked experiment and blocked experiment is RE = 20, indicating that blocking allows a massive reduction in sample size while keeping the same precision and power.

As expected, the t-test equally profits from the reduced standard error. The t-statistic is now $t = -2.9$ leading to a p-value of $p = 0.018$ and thus a significant result at the 5% significance level. This compares to the previous t-value of $t = -0.46$ for the unblocked design with a p-value of 0.65.

3.4 Sample Size and Precision

"How many samples do I need?" is arguably among the first questions a researcher asks when thinking about an experimental design. *Sample size determination* is a crucial component of experimental design in order to ensure that estimates are sufficiently precise to be of practical value and that hypothesis tests are adequately powered to be able to detect a relevant effect size. Sample size determination crucially depends on deciding on a minimal effect size. While precision and power can always be increased indefinitely by increasing the sample size, limits on resources—time, money, and available experimental material—pose practical limits. There is also a diminishing return, as doubling precision requires quadrupling the sample size.

3.4.1 Sample Size for Desired Precision

To provide a concrete example, let us consider our comparison of the two preparation kits again and assume that the volume of blood required is prohibitive for more than one sample per mouse. In the two-vendor experiment based on 20 mice, we found that our estimate $\hat{\Delta}$ was too imprecise to determine with any confidence which—if any—of the two kits yields lower responses than the other.

To determine a sufficient sample size, we need to decide which *minimal effect size* is relevant for us, a question answerable only with experience and subject-matter knowledge. For the sake of the example, let us say that a difference of $\delta_0 = \pm 0.5$ or larger would mean that we stick with one vendor, but a smaller difference is not of practical relevance for us. The task is therefore to determine the number n of mice per treatment group, such that the confidence interval of Δ has width no more than one, i.e., that

$$
\begin{aligned}
\text{UCL} - \text{LCL} &= (\hat{\Delta} + t_{1-\alpha/2,2n-2} \cdot \text{se}(\hat{\Delta})) - (\hat{\Delta} + t_{\alpha/2,2n-2} \cdot \text{se}(\hat{\Delta})) \\
&= (t_{1-\alpha/2,2n-2} - t_{\alpha/2,2n-2}) \cdot \sqrt{2}\sigma/\sqrt{n} \\
&\leq 2|\delta_0| \, .
\end{aligned}
$$

We note that the t-quantiles and the standard error both depend on n, which prevents us from solving this inequality directly. For a precise calculation, we can start at some not too large n, calculate the width of the confidence interval, increase n if the width is too large, and repeat until the desired precision is achieved.

If we have a reason to believe that n will not be very small, then we can reduce the problem to

$$
\text{UCL} - \text{LCL} = (z_{1-\alpha/2} - z_{\alpha/2})\sqrt{2}\sigma/\sqrt{n} \leq 2|\delta_0| \implies n \geq 2 \cdot z_{1-\alpha/2}^2 \sigma^2/\delta_0^2 \, ,
$$

if we exploit the fact that the t-quantile $t_{\alpha,n}$ is approximately equal to the standard normal quantile z_α, which does not depend on the sample size.

For a 95%-confidence interval, we have $z_{0.975} = +1.96$, which we can approximate as $z_{0.975} \approx 2$ without introducing any meaningful error. This leads to the simple formula

$$
n \geq 8\sigma^2/\delta_0^2 \, .
$$

In order to actually calculate the sample size with this formula, we need to know the standard deviation of the enzyme levels or an estimate $\hat{\sigma}$. Such an estimate might be available from previous experiments on the same problem. If not, we have to conduct a separate (usually small) experiment using a single treatment group for getting such an estimate. In our case, we already have an estimate $\hat{\sigma} = 1.65$ from our previous experiment, from which we find that a sample size of $n = 84$ mice *per kit* is required to reach our desired precision (the approximation $z_{0.975} = 2$ yields $n = 87$). This is a substantial increase in experimental material needed. We will have to decide if an experiment of this size is feasible for us, but a smaller experiment will likely waste time and resources without providing a practically relevant answer.

3.4.2 Precision for Given Sample Size

It is often useful to turn the question around: given we can afford a certain maximal number of mice for our experiment, what precision can we expect? If this precision turns out to be insufficient for our needs, we might as well call the experiment off or start considering alternatives.

For example, let us assume that we have 40 mice at our disposal for the experiment. From our previous discussion, we know that the variances of measurements using kits A and B can be assumed equal, so a balanced assignment of $n = 20$ mice per vendor is optimal. The expected width of a 95%-confidence interval is

$$\text{UCL} - \text{LCL} = (z_{0.975} - z_{0.025}) \cdot \sqrt{2}\sigma/\sqrt{n} = 1.24 \cdot \sigma .$$

Using our previous estimate of $\hat{\sigma} = 1.65$, we find the expected width of the 95%-confidence interval of 2.05 compared to 2.9 for the previous experiment with 10 mice per vendor, a decrease in length by $\sqrt{2} = 1.41$ due to doubling the sample size. This is not even close to our desired length of one, and we should consider if this experiment is worth doing, since it uses resources without a reasonable chance of providing a precise-enough estimate.

3.5 Sample Size and Power

A more common approach for determining the required sample size is via a hypothesis testing framework which allows us to also consider acceptable false positive and false negative probabilities for our experiment. For any hypothesis test, we can calculate each of the following five parameters from the other four:

- The **sample size** n.
- The **minimal effect size** δ_0 we want to reliably detect: the smaller the minimal effect size, the more samples are required to distinguish it from a zero effect. It is determined by subject-matter considerations and both raw and standardized effect sizes can be used.
- The **variance** σ^2: the larger the variance of the responses, the more samples we need to average out random fluctuations and achieve the necessary precision. Reducing the variance using the measures discussed above always helps, and blocking in particular is a powerful design option if available.
- The **significance level** α: the more stringent this level, the more samples we need to reliably detect a given effect. In practice, values of 5% or 1% are common.

- The **power** $1 - \beta$: this probability is higher the more stringent our α is set (for a fixed difference) and larger effects will lead to fewer false negatives for the same α level. In practice, the desired power is often about 80–90%; higher power might require prohibitively large sample sizes.

3.5.1 Power Analysis for Known Variance

We start developing the main ideas for determining a required sample size in a simplified scenario, where we know the variance exactly. Then, the standard error of $\hat{\Delta}$ is also known exactly, and the test statistic $\hat{\Delta}/\text{se}(\hat{\Delta})$ has a standard normal distribution under the null hypothesis $H_0 : \Delta = 0$. The same calculations can also be used with a variance estimate, provided the sample size is not too small and the t-distribution of the test statistic is well approximated by the normal distribution.

In the following, we assume that we decided on the false positive probability α, the power $1 - \beta$, and the minimal effect size $\Delta = \delta_0$. If the true difference is smaller than δ_0, we might still detect it, but detection becomes less and less likely the smaller the difference gets. If the difference is greater, our chance of detection increases.

If $H_0 : \Delta = 0$ is true, then $\hat{\Delta} \sim N(0, 2\sigma^2/n)$ has a normal distribution with mean zero and variance $2\sigma^2/n$. We reject the null hypothesis if $\hat{\Delta} \leq z_{\alpha/2} \cdot \sqrt{2}\sigma/\sqrt{n}$ or $\hat{\Delta} \geq z_{1-\alpha/2} \cdot \sqrt{2}\sigma/\sqrt{n}$. These two critical values are shown as dashed vertical lines in Fig. 3.1 (top) for sample sizes $n = 10$ (left) and $n = 90$ (right). As expected, the critical values move closer to zero with increasing sample size.

If H_0 is not true and $\Delta = \delta_0$ instead, then $\hat{\Delta} \sim N(\delta_0, 2\sigma^2/n)$ has a normal distribution with mean δ_0 and variance $2\sigma^2/n$. This distribution is shown in the bottom row of Fig. 3.1 for the two sample sizes and a true difference of $\delta_0 = 1$; it also gets narrower with increasing sample size n.

A false negative happens if H_0 is not true, so $\Delta = \delta_0$, yet the estimator $\hat{\Delta}$ falls outside the rejection region. The probability of this event is

$$ \mathbb{P}\left(|\hat{\Delta}| \leq z_{1-\alpha/2} \cdot \frac{\sqrt{2} \cdot \sigma}{\sqrt{n}}; \Delta = \delta_0 \right) = \beta , $$

which yields the fundamental equality

$$ z_{1-\alpha/2} \cdot \sqrt{2}\sigma/\sqrt{n} = z_\beta \cdot \sqrt{2}\sigma/\sqrt{n} + \delta_0 . \tag{3.1} $$

Our goal is to find n such that the probability of a false negative stays below a prescribed value β.

We can see this in Fig. 3.1: for a given $\alpha = 5\%$, the dashed lines denote the rejection region, and the black shaded area corresponds to a probability of 5%. If $\Delta = \delta_0$, we get the distributions in the bottom row, where all values inside the dashed lines are false negatives, and the probability β corresponds to the gray shaded area.

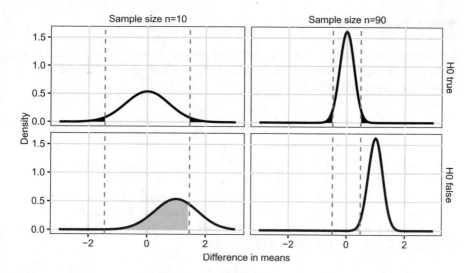

Fig. 3.1 Distributions of difference in means if the null hypothesis is true and the difference between means is zero (top) and when the alternative hypothesis is true and the difference is one (bottom) for 10 (left) and 90 (right) samples. The dashed lines are the critical values for the test statistic. Shaded black region: false positives (α). Shaded gray region: false negatives (β)

Increasing n from $n = 10$ (left) to $n = 90$ (right) narrows both distributions, moves the critical values toward zero, and shrinks the false negative probability.

We find a closed formula for the sample size by solving Eq. (3.1) for n:

$$n = \frac{2 \cdot (z_{1-\alpha/2} + z_{1-\beta})^2}{(\delta_0/\sigma)^2} = \frac{2 \cdot (z_{1-\alpha/2} + z_{1-\beta})^2}{d_0^2}, \qquad (3.2)$$

where we used the fact that $z_\beta = -z_{1-\beta}$. The first formula uses the minimal raw effect size δ_0 and requires knowledge of the residual variance, whereas the second formula is based on the minimal standardized effect size $d_0 = \delta_0/\sigma$, which measures the difference between the means as a multiple of the standard deviation.

In our example, a hypothesis test with significance level $\alpha = 5\%$ and a variance of $\sigma^2 = 2$ has power 11, 35, and 100% to detect a true difference of $\delta_0 = 1$ based on $n = 2$, $n = 10$, and $n = 100$ mice per treatment group, respectively. We require at least 31 mice per vendor to achieve a power of $1 - \beta = 80\%$.

The same ideas apply to calculating the minimal effect size that is detectable with a given significance level and power for any fixed sample size. For our example, we might only have 20 mice per vendor at our disposal. For our variance of $\sigma^2 = 2$, a significance level of $\alpha = 5\%$ and a power of $1 - \beta = 80\%$, we find that for $n = 20$, the achievable minimal effect size is $\delta_0 = 1.25$.

A small minimal standardized effect size of $d_0 = \delta_0/\sigma = 0.2$ requires at least $n = 392$ mice per vendor for $\alpha = 5\%$ and $1 - \beta = 80\%$. This number decreases to $n = 63$ and $n = 25$ for a medium effect $d_0 = 0.5$, respectively, a larger effect $d_0 = 0.8$.

Portable Power Calculation

It is convenient to have simple approximate formulas to find a rough estimate of the required sample size for a desired comparison. For our two-sample problem, an approximate power formula is

$$
n \approx \frac{16}{(\delta_0/\sigma)^2} = \frac{16}{d_0^2} \, ,
\tag{3.3}
$$

based on the observation that the numerator in Eq. (3.2) is then roughly 16 for a significance level $\alpha = 5\%$ and a reasonable power of $1 - \beta = 80\%$.

Such approximate formulas were termed *portable power* (Wheeler 1974) and enable quick back-of-napkin calculation during a discussion, for example.

We can translate the sample size formula to a *relative* effect based on the coefficient of variation $CV = \sigma/\mu$ (Belle and Martin 1993):

$$
n \approx \frac{16 \cdot (CV)^2}{\ln(\mu_A/\mu_B)^2} \, .
$$

This requires taking logarithms and is not quite so portable. A convenient further shortcut exists for a variation of 35%, typical for biological systems (Belle 2008), noting that the numerator then simplifies to $16 \cdot (0.35)^2 \approx 2$.

For example, a difference in enzyme level of at least 20% of vendor A compared to vendor B and a variability for both vendors of about 30% means that

$$
\frac{\mu_A}{\mu_B} = 0.8 \quad \text{and} \quad \frac{\sigma_A}{\mu_A} = \frac{\sigma_B}{\mu_B} = 0.3 \, .
$$

The necessary sample size per vendor is then

$$
n \approx \frac{16 \cdot (0.3)^2}{\ln(0.8)^2} \approx 29 \, .
$$

For a higher variability of 35%, the sample size increases, and our shortcut yields $n \approx 2/\ln(\mu_A/\mu_B)^2 \approx 40$.

3.5.2 Power Analysis for Unknown Variance

In practice, the variance σ^2 is usually not known and the test statistic T uses an estimate $\hat{\sigma}^2$ instead. If H_0 is true and $\Delta = 0$, then T has a t-distribution with $2n - 2$ degrees of freedom, and its quantiles depend on the sample size. If H_0 is false and the true difference is $\Delta = \delta_0$, then the test statistic has a noncentral t-distribution with noncentrality parameter

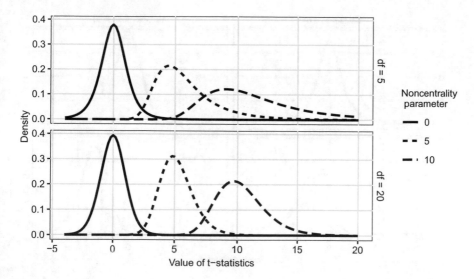

Fig. 3.2 t-distribution for 5 (top) and 20 (bottom) degrees of freedom and three different noncentrality parameters (linetype)

$$\eta = \frac{\delta_0}{\text{se}(\hat{\Delta})} = \frac{\delta_0}{2 \cdot \sigma/\sqrt{n}} = \sqrt{n/2} \cdot \frac{\delta_0}{\sigma} = \sqrt{n/2} \cdot d_0 .$$

For illustration, Fig. 3.2 shows the density of the t-distribution for different number of samples and different values of the noncentrality parameter; note that the noncentral t-distribution is not symmetric, and $t_{\alpha,n}(\eta) \neq -t_{1-\alpha,n}(\eta)$. The noncentrality parameter can be written as $\eta^2 = 2 \cdot n \cdot (d_0^2/4)$, the product of the experiment size $2n$, and the (squared) effect size.

An increase in sample size has two effects on the distribution of the test statistic T: (i) it moves the critical values inwards, although this effect is only pronounced for small sample sizes; (ii) it increases the noncentrality parameter with $\sqrt{n}$ and thereby shifts the distribution away from zero. This is in contrast to our earlier discussion using a normal distribution, where an increase in sample size results in a decrease in the variance.

In other words, increasing the sample size slightly alters the shape of the distribution of our test statistic, but more importantly moves it away from the central t-distribution under the null hypothesis. The overlap between the two distributions then decreases and the same observed difference between the two treatment means is more easily distinguished from a zero difference.

As an example, assume we have a reason to believe that the two kits indeed give consistently different readouts. For a significance level of $\alpha = 5\%$ and $n = 10$, we calculate the power that we successfully detect a true difference of $|\Delta| = \delta_0 = 2$, of $\delta_0 = 1$, and of $\delta_0 = 0.5$. Under the null hypothesis, the test statistics T has a (central) t-distribution with $2n - 2$ degrees of freedom, and we reject the hypothesis

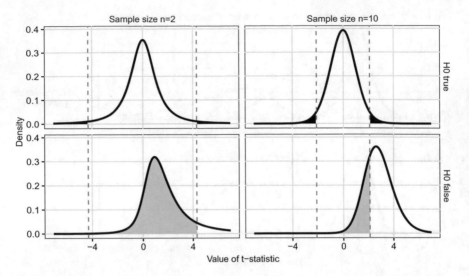

Fig. 3.3 Distributions of t-statistic if null hypothesis is true and true difference is zero (top) and when alternative hypothesis is true and true difference is $\delta_0 = 2$ (bottom) for 2 (left) and 10 (right) samples. The dashed lines are the critical values for the test statistic. Shaded black region: false positives (α). Shaded gray region: false negatives β

if $|T| > t_{1-\alpha/2, 2n-2}$ (Fig. 3.3 (top)). If however, $\Delta = \delta_0$ is true, then the distribution of T changes to a noncentral t-distribution with $2n - 2$ degrees of freedom and noncentrality parameter $\eta = \delta_0/\mathrm{se}(\hat{\Delta})$ (Fig. 3.3 (bottom)). The power $1 - \beta$ is the probability that this T falls into the rejection region and either stays above $t_{1-\alpha/2, 2n-2}$ or below $t_{\alpha/2, 2n-2}$.

We compute the upper t-quantile for $n = 10$ as $t_{0.975, 18} = 2.1$. If the true difference is $\delta_0 = 2$, then the probability to stay above this value (and correctly reject H_0) is high with a power of 73%. This is because the standard error is 0.74 and thus the precision of the estimate $\hat{\Delta}$ is large compared to the difference we attempt to detect. Decreasing this difference while keeping the significance level and the sample size fixed decreases the power to 25% for $\delta_0 = 1$ and further to 10% for $\delta_0 = 0.5$. In other words, we can expect to detect a true difference of $\delta_0 = 0.5$ in only 10% of experiments with 10 samples per vendor and it is questionable if such an experiment is worth implementing.

It is not possible to find a closed formula for the sample size calculation, because the central and noncentral t-quantiles depend on n, while the noncentrality parameter depends on n and additionally alters the shape of the noncentral t-distribution. R's built-in function `power.t.test()` uses an iterative approach and yields a (deceptively precise!) sample size of $n = 43.8549046$ per vendor to detect a difference $\delta_0 = 1$ with 80% power at a 5% significance level, based on our previous estimate $\hat{\sigma}^2 = 2.73$ of the variance. We provide an iterative algorithm for illustration in Sect. 3.6.

Note that we replaced the unknown true variance with an estimate $\hat{\sigma}^2$ for this calculation, and the accuracy of the resulting sample size hinges partly on the assumption that the variance estimate is reasonably close to the true variance.

3.5.3 Power Analysis in Practice

Power calculations are predicated on multiple assumptions such as normally distributed response values and independence of observations, all of which might hold only approximately. They also require a value for the residual variance, which is usually not known before the experiment.

In practice, we therefore base our calculations on an educated guess of this variance, or on an estimate from a previous experiment. The residual variance can sometimes be estimated from a *pilot experiment*, conducted to 'test-run' the experimental conditions and protocols of the anticipated experiment on a smaller scale. Alternatively, small experiments with a single treatment group (usually the control group) can also provide an estimate, as can the perusal of previous experiments with similar experimental material.

Even though methods like `power.t.test()` return the sample size n with deceptively precise six decimals, it is prudent to consider the uncertainty in the variance estimate and its impact on the resulting power and sample size and to allow for a margin of error.

A very conservative approach is to consider a worst-case scenario and use the upper confidence limit instead of the variance estimate for the power analysis. In our example, we have a variance estimate of $\hat{\sigma}^2 = 2.73$. The 95%-confidence interval of this estimate is [1.56, 5.97] and is based on $2n - 2 = 18$ degrees of freedom. For a desired power of 80% and a significance level of 5%, we found that we need 44 mice per group for detecting a difference of $\delta_0 = 1$ based on the point estimate for σ^2. If we use the upper confidence limit instead, the sample size increases by $\text{UCL}/\hat{\sigma}^2 \approx 5.97 / 2.73 = 2.2$ to $n = 95$.

A more optimistic approach uses the estimated variance for the calculation and then adds a safety margin to the resulting sample size to compensate for uncertainties. The margin is fully at the discretion of the researcher, but adding 20%–30% to the sample size seems to be reasonable in many cases. This increases the sample size in our example from 44 to 53 for a 20% margin, and to 57 for a 30% margin.

3.5.4 'Observed Power' and Related Fallacies

When confronted with an undesired non-significant test outcome from their experiment, researchers sometimes calculate the *observed power* or *retrospective power* based on the effect size and residual variance estimated from the data. It is then argued that large observed power provides evidence in favor of the null hypothesis. Appealing as this idea might seem, it is fundamentally flawed and based on the improper use of the concept of power.

To see this, let us imagine that two t-tests are performed, with resulting test statistics t_1 and t_2, and associated non-significant p-values $p_1 > \alpha$ and $p_2 > \alpha$. Assume that $t_1 > t_2$ and the first hypothesis test indicates a larger deviation from the null hypothesis. The first p-value is then smaller than the second, which we interpret as stronger—yet not significant—evidence against the null hypothesis in the first test. If we now calculate the observed power at t_1, respectively t_2, for our desired significance level α, we find that this power is larger for the first experiment since t_1 is further away from the zero value. The proposed argument then claims that this larger observed power provides more evidence in favor of the null hypothesis. This directly contradicts our previous interpretation of the two p-values.

The fallacy arises because p-value and observed power are both based on the same (random) values of the test statistic and residual variance estimated from the specific data of each experiment. This always results in higher observed power for lower p-values and leads to the apparent contradiction in the example. Indeed, it can be shown that the observed power is in direct correspondence to the p-value and therefore provides no additional information.

Similar problems result if the power is based on the observed residual variance, but calculated at a specific effect size deemed scientifically relevant. Because it is again based on the specific outcome of the experiment, this power cannot be interpreted as the power to detect an effect of the specific size and provides no evidence in favor or against the null hypothesis.

We have already seen a much better—and logically correct—way of interpreting a (non-significant) test result by estimating the difference and calculating its confidence interval. In contrast to observed power, p-value and confidence interval provide different pieces of information and there is no direct correspondence between. If the interval is wide and contains the value zero, as in our unpaired-vendor example, we conclude that the data provide little evidence for or against the null hypothesis. If the interval is short, as in our paired-vendor example, we conclude that plausible values are restricted to a narrow range. If this range includes zero, we have evidence that the true value is unlikely to be far off zero.

An equivalent argument can be made using two one-sided significance tests, where the null hypotheses are $H_0 : \Delta < -\delta_0$ and $H_0 : \Delta > +\delta_0$. Note that these reverse the burden of proof and a rejection of both hypotheses means that the true difference is likely in the interval $(-\delta_0, +\delta_0)$. This is known as an *equivalence test*, where the aim is to show that two treatments are equal (rather than different). These tests play a prominent role in toxicity or environmental studies, where we try to demonstrate that responses from a treatment group exposed to a potential hazard are no different than $\pm\delta_0$ compared to a non-exposed group.

Calculating power *prospectively* in the planning of an experiment ensures that tests are adequately powered and estimates are sufficiently precise. Power analysis has no role *retrospectively* in the analysis of data from an experiment. Here, estimates of effect sizes and their confidence intervals are the appropriate quantities, augmented by p-values if necessary.

3.6 Notes and Summary

Notes
A general discussion of power and sample size is given in Cohen (1992, 1988) and
Krzywinski and Altman (2013), and a gentle practical introduction is Lenth (2001).
Sample sizes for confidence intervals are addressed in Goodman and Berlin (1994),
Maxwell et al. (2008), and Rothman and Greenland (2018); the equivalence to power
calculation based on testing is discussed in Altman et al. (2000). The fallacies of
'observed power' are elucidated in Hoenig and Heisey (2001) and equivalence testing
in Schuirmann (1987). The free software G*Power is an alternative to R for sample
size determination (Faul 2007).

Power analysis code
To further illustrate our power analysis, we implement the corresponding calculations
in an R function. The following code calculates the power of a t-test given the minimal
difference delta, the significance level alpha, the sample size n, and the standard
deviation s.

The function first calculates the degrees of freedom for the given sample size.
Then, the lower and upper critical values for the t-statistic under the null hypothesis
are computed. Next, the noncentrality parameter $\eta = $ ncp is used to determine the
probability of correctly rejecting H_0 if indeed $|\Delta| = \delta_0$; this is precisely the power.

We start from a reasonably low n, calculate the power, and increase the sample
size until the desired power is reached.

```
# delta0: true difference
# alpha: significance level (false positive rate)
# n: sample size
# s: standard deviation
# return: power
getPowerT = function(delta0, alpha, n, s) {
  df = 2*n-2   # degrees of freedom
  q.H0.low = qt(p=alpha/2, df=df)     # low rejection quantile
  q.H0.high = qt(p=1-alpha/2, df=df) # high rejection quantile
  ncp = abs(delta0) / (sqrt(2)*s/sqrt(n)) # noncentrality
  # prob. to reject low or high values if H0 false
  p.low = pt(q=q.H0.low, df=df, ncp=ncp)
  p.high = 1 - pt(q=q.H0.high, df=df, ncp=ncp)
  return( p.low + p.high )
}
```

Using R
Base-R provides the power.t.test() function for power calculations based on
the t-distribution. It takes four of the five parameters and calculates the fifth.

Summary
Determining the required sample size of an experiment—at least approximately—is
part of the experimental design. We can use the hypothesis testing framework to
determine sample size based on the two error probabilities, a measure of variation,

and the required minimal effect size. The resulting sample size should then be used to determine if estimates have sufficient expected precision. We can also determine the minimal effect size detectable with desired power and sample size, or the power achieved from a given sample size for a minimal effect size, all of which we can use to decide if an experiment is worth doing. Precision and power can also be increased without increasing the sample size, by balanced allocation, narrowing experimental conditions, or blocking. Power analysis is often based on noncentral distributions, whose noncentrality parameters are the product of experiment size and effect size; portable power formulas use approximations of various quantities to allow back-of-envelope power analysis.

References

Altman, D. G. et al. (2000). Statistics with Confidence. 2nd. John Wiley & Sons, Inc.

Cohen, J. (1988). Statistical Power Analysis for the Behavioral Sciences. 2nd. Lawrence Erlbaum Associates, Hillsdale.

Cohen, J. (1992). "A Power Primer". In: Psychological Bulletin 112.1, pp. 155–159.

Faul, F. et al. (2007). "G∗Power3: a flexible statistical power analysis program for the social, behavioral, and biomedical sciences". In: Behavior Research Methods 39.2, pp. 175–191.

Goodman, S. N. and J. A. Berlin (1994). "The Use of Predicted Confidence Intervals When Planning Experiments and the Misuse of Power When Interpreting Results". In: Annals of Internal Medicine 121, pp. 200–206.

Hoenig, J. M. and D. M. Heisey (2001). "The abuse of power: The pervasive fallacy of power calculations for data analysis". In: American Statistician 55.1, pp. 19–24.

Hurlbert, S. H. (1984). "Pseudoreplication and the Design of Ecological Field Experiments". In: Ecological Monographs 54.2, pp. 187–211.

Hurlbert, S. H. (2009). "The ancient black art and transdisciplinary extent of pseudoreplication". In: Journal of Comparative Psychology 123.4, pp. 434–443.

Krzywinski, M. and N. Altman (2013). "Points of significance: Power and sample size". In: Nature Methods 10, pp. 1139–1140.

Lenth, R. V. (2001). "Some Practical Guidelines for Effective Sample Size Determination". In: The American Statistician 55.3, pp. 187–193.

Maxwell, S. E., K. Kelley, and J. R. Rausch (2008). "Sample Size Planning for Statistical Power and Accuracy in Parameter Estimation". In: Annual Review of Psychology 59.1, pp. 537–563.

Rothman, K. J. and S. Greenland (2018). "'Planning Study Size Based on Precision Rather Than Power". In: Epidemiology 29.5, pp. 599–603.

Schuirmann, D. J. (1987). "A comparison of the two one-sided tests procedure and the power approach for assessing the equivalence of average bioavailability". In: Pharmacometrics 15.6, pp. 657–680.

van Belle, G. (2008). Statistical Rules of Thumb. 2nd. John Wiley & Sons, Inc.

van Belle, G. and D. C. Martin (1993). "Sample size as a function of coefficient of variation and ratio". In: American Statistician 47.3, pp. 165–167.

Wheeler, R. E. (1974). "Portable Power". In: Technometrics 16.2, pp. 193–201.

Chapter 4
Comparing More Than Two Groups: One-Way ANOVA

4.1 Introduction

We extend our discussion from experiments with two treatment groups to experiments with k treatment groups, assuming completely random treatment allocation. In this chapter, we develop the *analysis of variance* framework to address the *omnibus null hypothesis* that all group means are equal and there is no difference between the treatments. The main idea is to partition the overall variation in the data into one part attributable to differences between the treatment group means, and a residual part. We can then test the equality of group means by comparing variances between group means and within each group using an F-test.

We also look at corresponding effect size measures to quantify the overall difference between treatment group means. The associated power analysis uses the noncentral F-distribution, where the noncentrality parameter is the product of experiment size and effect size. Several simplifications allow us to derive a portable power formula for quickly approximating the required sample size.

The analysis of variance is intimately linked to a linear model, and we formally introduce Hasse diagrams to describe the logic of the experiment and to derive the corresponding linear model and its specification in statistical software.

4.2 Experiment and Data

We consider investigating four drugs for their properties to alter the metabolism in mice, and we take the level of a liver enzyme as a biomarker to indicate this alteration, where higher levels are considered 'better'. Metabolization and elimination of the drugs might be affected by the fatty acid metabolism, but for the moment we control this aspect by feeding all mice with the same low-fat diet and return to the diet effect in Chap. 6.

© Springer Nature Switzerland AG 2021
H.-M. Kaltenbach, *Statistical Design and Analysis of Biological Experiments*,
Statistics for Biology and Health, https://doi.org/10.1007/978-3-030-69641-2_4

Table 4.1   Measured enzyme levels for four drugs assigned to eight mice each

D1	13.94	16.04	12.70	13.98	14.31	14.71	16.36	12.58
D2	13.56	10.88	14.75	13.05	11.53	13.22	11.52	13.99
D3	10.57	8.40	7.64	8.97	9.38	9.13	9.81	10.02
D4	10.78	10.06	9.29	9.36	9.04	9.41	6.86	9.89

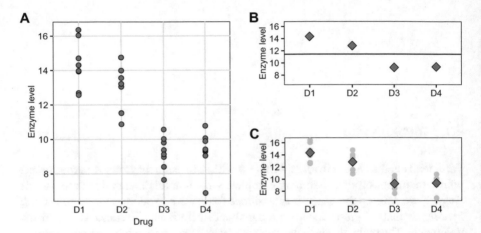

Fig. 4.1   Observed enzyme levels in response to four drug treatments with eight mice per treatment group. **A** Individual observations by treatment group. **B** Grand mean (horizontal line) and group means (diamonds) used in estimating the between-groups variance. **C** Individual responses (gray points) are compared to group means (black diamonds) for estimating the within-group variance

The data in Table 4.1 and Fig. 4.1A show the observed enzyme levels for $N = n \cdot k = 32$ mice, with $n = 8$ mice randomly assigned to one of the $k = 4$ drugs $D1$, $D2$, $D3$, and $D4$. We denote the four average treatment group responses as $\mu_1, \ldots, \mu_4$; we are interested in testing the omnibus hypothesis $H_0 : \mu_1 = \mu_2 = \mu_3 = \mu_4$ that the group averages are identical and the four drugs therefore all have the same effect on the enzyme levels.

Other interesting questions regard the estimation and testing of specific treatment group comparisons, which we postpone to Chap. 5.

4.3   One-Way Analysis of Variance

In a balanced *completely randomized design*, we randomly allocate k treatments on $N = n \cdot k$ experimental units. We assume that the response $y_{ij} \sim N(\mu_i, \sigma_e^2)$ of the jth experimental unit in the ith treatment group is normally distributed with group-specific expectation μ_i and common variance σ_e^2, with $i = 1 \ldots k$ and $j = 1 \ldots n$; each group then has n experimental units.

4.3.1 Testing Equality of Means by Comparing Variances

For $k = 2$ treatment groups, the omnibus null hypothesis is $H_0 : \mu_1 = \mu_2$ and can be tested using a t-test on the group difference $\Delta = \mu_1 - \mu_2$. For $k > 2$ treatment groups, the corresponding omnibus null hypothesis is $H_0 : \mu_1 = \cdots = \mu_k$, and the idea of using a single difference for testing does not work.

The crucial observation for deriving a test statistic for the general omnibus null hypothesis comes from changing our perspective on the problem: if the treatment group means $\mu_i \equiv \mu$ are all equal, then we can consider their estimates $\hat{\mu}_i = \sum_{j=1}^{n} y_{ij}/n$ as independent 'samples' from a normal distribution with mean μ and variance σ_e^2/n (Fig. 4.1B). We can then estimate their variance using the usual formula

$$\widehat{\mathrm{Var}}(\hat{\mu}_i) = \sum_{i=1}^{k} (\hat{\mu}_i - \hat{\mu})^2/(k-1) \,,$$

where $\hat{\mu} = \sum_{i=1}^{k} \hat{\mu}_i/k$ is an estimate of the grand mean μ. Since $\mathrm{Var}(\hat{\mu}_i) = \sigma_e^2/n$, this provides us with an estimator

$$\tilde{\sigma}_e^2 = n \cdot \widehat{\mathrm{Var}}(\hat{\mu}_i) = n \cdot \sum_{i=1}^{k} (\hat{\mu}_i - \hat{\mu})^2/(k-1)$$

for the variance σ_e^2 that only considers the dispersion of group means around the grand mean and is independent of the dispersion of individual observations around their group mean.

On the other hand, our previous estimator pooled over groups is

$$\hat{\sigma}_e^2 = \left(\underbrace{\frac{\sum_{j=1}^{n}(y_{1j} - \hat{\mu}_1)^2}{n-1}}_{\text{variance group 1}} + \cdots + \underbrace{\frac{\sum_{j=1}^{n}(y_{kj} - \hat{\mu}_k)^2}{n-1}}_{\text{variance group k}} \right) \bigg/ k = \sum_{i=1}^{k}\sum_{j=1}^{n} \frac{(y_{ij} - \hat{\mu}_i)^2}{N-k}$$

and also estimates the variance σ_e^2 (Fig. 4.1C). It only considers the dispersion of observations around their group means and is independent of the μ_i being equal. For example, we could add a fixed number to all measurements in one group and this would affect $\tilde{\sigma}_e^2$ but not $\hat{\sigma}_e^2$.

The two estimators have expectations

$$\mathbb{E}(\tilde{\sigma}_e^2) = \sigma_e^2 + \underbrace{\frac{1}{k-1} \sum_{i=1}^{k} n \cdot (\mu_i - \mu)^2}_{\varrho} \quad \text{and} \quad \mathbb{E}(\hat{\sigma}_e^2) = \sigma_e^2 \,,$$

respectively. Thus, while $\hat{\sigma}_e^2$ is always an unbiased estimator of the residual variance, the estimator $\tilde{\sigma}_e^2$ has bias Q—the *between-groups variance*—but is unbiased
precisely if H_0 is true and $Q = 0$. If H_0 is true, then the ratio

$$F = \frac{\tilde{\sigma}_e^2}{\hat{\sigma}_e^2} \sim F_{k-1,N-k}$$

has an F-distribution with $k - 1$ numerator and $N - k$ denominator degrees of freedom. The distribution of F deviates more from this distribution the larger the term
Q becomes and the 'further away' the observed group means are from equality.
The corresponding F-test thus provides the required test of the omnibus hypothesis
$H_0 : \mu_1 = \cdots = \mu_k$.

For our example, we estimate the four group means $\hat{\mu}_1 = 14.33$, $\hat{\mu}_2 = 12.81$,
$\hat{\mu}_3 = 9.24$ and $\hat{\mu}_4 = 9.34$, and a grand mean $\hat{\mu} = 11.43$. Their squared differences
are $(\hat{\mu}_1 - \hat{\mu})^2 = 8.42$, $(\hat{\mu}_2 - \hat{\mu})^2 = 1.91$, $(\hat{\mu}_3 - \hat{\mu})^2 = 4.79$, and $(\hat{\mu}_4 - \hat{\mu})^2 = 4.36$.
With $n = 8$ mice per treatment group, this gives the two estimates $\tilde{\sigma}_e^2 = 51.96$ and
$\hat{\sigma}_e^2 = 1.48$ for the variance. Their ratio is $F = 35.17$ and comparing it to an F-
distribution with $k - 1 = 3$ numerator and $N - k = 28$ denominator degrees of freedom gives a p-value of $p = 1.24 \times 10^{-9}$, indicating strongly that the group means
are not identical.

Perhaps surprisingly, the F-test is identical to the t-test for $k = 2$ treatment groups,
and will give the same p-value. We can see this directly using a little algebra (note
$\hat{\mu} = (\hat{\mu}_1 + \hat{\mu}_2)/2$ and $N = 2 \cdot n$):

$$F = \frac{n \cdot ((\hat{\mu}_1 - \hat{\mu})^2 + (\hat{\mu}_2 - \hat{\mu})^2)/1}{\hat{\sigma}_e^2} = \frac{\frac{n}{2} \cdot (\hat{\mu}_1 - \hat{\mu}_2)^2}{\hat{\sigma}_e^2} = \left(\frac{\hat{\mu}_1 - \hat{\mu}_2}{\sqrt{2}\hat{\sigma}_e/\sqrt{n}} \right)^2 = T^2 ,$$

and $F_{1,2n-2} = t_{2n-2}^2$. While the t-test uses a difference in means, the F-test uses the
average squared difference from group means to general mean; for two groups, these
concepts are equivalent.

4.3.2 Analysis of Variance

Our derivation of the omnibus F-test used the decomposition of the data into a
between-groups and a within-groups component. We can exploit this decomposition
further in the *(one-way) analysis of variance (ANOVA)* by directly partitioning the
overall variation in the data via *sums of squares* and their associated *degrees of
freedom*. In the words of its originator:

> **The analysis of variance is not a mathematical theorem, but rather a convenient method
> of arranging the arithmetic.** (Fisher 1934)

The arithmetic advantages of the analysis of variance are no longer relevant today, but the decomposition of the data into various parts for explaining the observed variation remains an easily interpretable summary of the experimental results.

Partitioning the Data

To stress that ANOVA decomposes the variation in the data, we first write each datum y_{ij} as a sum of three components: the grand mean, deviation of group mean to grand mean, and deviation of datum to group mean:

$$y_{ij} = \bar{y}.. + (\bar{y}_{i.} - \bar{y}..) + (y_{ij} - \bar{y}_{i.}) ,$$

where $\bar{y}_{i.} = \sum_j y_{ij}/n$ is the average of group i, $\bar{y}.. = \sum_i \sum_j y_{ij}/nk$ is the grand mean, and a dot indicates summation over the corresponding index.

For example, the first datum in the second group, $y_{21} = 13.56$, is decomposed into the grand mean $\bar{y}.. = 11.43$, the deviation from group mean $\bar{y}_{2.} = 12.81$ to grand mean $(\bar{y}_2. - \bar{y}..) = 1.38$, and the residual $y_{21} - \bar{y}_2. = 0.75$.

Sums of Squares

We quantify the overall variation in the observations by the *total sum of squares*, the summed squared distances of each datum y_{ij} to the estimated grand mean $\bar{y}...$

Following the partition of each datum, the total sum of squares is also partitioned into two parts: (i) the *treatment (or between-groups) sum of squares* which measures the variation between group means and captures the variation explained by the systematic differences between the treatments, and (ii) the *residual (or within-groups) sum of squares* which measures the variation of responses within each group and thus captures the unexplained random variation:

$$\text{SS}_{\text{tot}} = \sum_{i=1}^{k}\sum_{j=1}^{n}(y_{ij} - \bar{y}..)^2 = \underbrace{\sum_{i=1}^{k} n \cdot (\bar{y}_{i.} - \bar{y}..)^2}_{\text{SS}_{\text{trt}}} + \underbrace{\sum_{i=1}^{k}\sum_{j=1}^{n}(y_{ij} - \bar{y}_{i.})^2}_{\text{SS}_{\text{res}}} .$$

The intermediate term $2\sum_i \sum_j (y_{ij} - \bar{y}_{i.})(\bar{y}_{i.} - \bar{y}..) = 0$ vanishes because SS_{trt} is based on group means and grand mean, while SS_{res} is independently based on observations and group means; the two are *orthogonal*.

For our example, we find a total sum of squares of $\text{SS}_{\text{tot}} = 197.26$, a treatment sum of squares $\text{SS}_{\text{trt}} = 155.89$, and a residual sum of squares $\text{SS}_{\text{res}} = 41.37$; as expected, the latter two add precisely to SS_{tot}. Thus, most of the observed variation in the data is due to systematic differences between the treatment groups.

Degrees of Freedom

Associated with each sum of squares term are its *degrees of freedom*, the number of independent components used to calculate it.

The total degrees of freedom for SS_{tot} are $df_{tot} = N - 1$, because we have N response values, and need to compute a single value $\bar{y}_{..}$ to find the sum of squares.

The treatment degrees of freedom are $df_{trt} = k - 1$, because there are k treatment means, estimated by $\bar{y}_{i.}$, but the calculation of the sum of squares requires the overall average $\bar{y}_{...}$.

Finally, there are N residuals, but we used up 1 degree of freedom for the overall average, and $k - 1$ for the group averages, leaving us with $df_{res} = N - k$ degrees of freedom.

The degrees of freedom then decompose as

$$df_{tot} = df_{trt} + df_{res} .$$

This decomposition tells us how much of the data we 'use up' for calculating each sum of squares component.

Mean Squares

Dividing a sum of squares by its degrees of freedom gives the corresponding *mean squares*, which are exactly our two variance estimates. The *treatment mean squares* are given by

$$MS_{trt} = \frac{SS_{trt}}{df_{trt}} = \frac{SS_{trt}}{k - 1} = \tilde{\sigma}_e^2$$

and are our first variance estimate based on group means and grand mean, while the *residual mean squares*

$$MS_{res} = \frac{SS_{res}}{df_{res}} = \frac{SS_{res}}{N - k} = \hat{\sigma}_e^2$$

are our second independent estimator for the within-group variance. We find $MS_{res} = 41.37/28 = 1.48$ and $MS_{trt} = 155.89/3 = 51.96$ for our example.

In contrast to the sum of squares, the mean squares do *not* decompose by factor and $MS_{tot} = SS_{tot}/(N - 1) = 6.36 \neq MS_{trt} + MS_{res} = 53.44$.

Omnibus F-Test

Our F-statistic for testing the omnibus hypothesis $H_0 : \mu_1 = \cdots = \mu_k$ is then

Table 4.2 Manually calculated summary table for analysis of variance

Source	df	SS	MS	F	p
Drug	3	155.89	51.96	35.2	1.24e-09
Residual	28	41.37	1.48		

$$F = \frac{MS_{trt}}{MS_{res}} = \frac{SS_{trt}/df_{trt}}{SS_{res}/df_{res}} \sim F_{df_{trt}, df_{res}} ,$$

and we reject H_0 if the observed F-statistic exceeds the $(1 - \alpha)$-quantile $F_{1-\alpha, df_{trt}, df_{res}}$.

Based on the sum of squares and degrees of freedom decompositions, we again find the observed test statistic of $F = 51.96/1.48 = 35.17$ on $df_{trt} = 3$ and $df_{res} = 28$ degrees of freedom, corresponding to a p-value of $p = 1.24 \times 10^{-9}$.

ANOVA Table

The results of an ANOVA are usually presented in an *ANOVA table* such as Table 4.2 for our example.

An ANOVA table has one row for each source of variation, and the first column gives the name of each source. The remaining columns give (i) the degrees of freedom available for calculating the sum of squares (indicating how much of the data is 'used' for this source of variation), (ii) the sum of squares to quantify the variation attributed to the source, (iii) the resulting mean squares used for testing, (iv) the observed value of the F-statistic for the omnibus null hypothesis, and (v) the corresponding p-value.

4.3.3 Effect Size Measures

The raw difference Δ or the standardized difference d are both easily interpretable effect size measures for the case of $k = 2$ treatment groups that we can use in conjunction with the t-test. We now introduce three effect size measures for the case of $k > 2$ treatment groups for use in conjunction with an omnibus F-test.

A simple effect size measure is the *variation explained*, which is the proportion of the factor's sum of squares of the total sum of squares:

$$\eta_{trt}^2 = \frac{SS_{trt}}{SS_{tot}} = \frac{SS_{trt}}{SS_{trt} + SS_{res}} .$$

A large value of η_{trt}^2 indicates that the majority of the variation is not due to random variation between observations in the same treatment group, but rather due to the fact that the average responses to the treatments are different. In our example, we find $\eta_{\text{trt}}^2 = 155.89/197.26 = 0.79$, confirming that the differences between drugs are responsible for 79% of the variation in the data.

The *raw effect size* measures the average deviation between group means and the grand mean:

$$b^2 = \frac{1}{k}\sum_{i=1}^{k}(\mu_i - \mu)^2 = \frac{1}{k}\sum_{i=1}^{k}\alpha_i^2 \, .$$

A corresponding *standardized effect size* was proposed in Cohen (1992):

$$f^2 = b^2/\sigma_e^2 = \frac{1}{k\sigma_e^2}\sum_{i=1}^{k}\alpha_i^2$$

and measures the average deviation between group means and grand mean in units of residual variance. It specializes to $f = d/2$ for $k = 2$ groups.

Extending from his classification of effect sizes d, Cohen proposed that values of $f > 0.1$, $f > 0.25$, and $f > 0.4$ may be interpreted as small, medium, and large effects (Cohen 1992). An unbiased estimate of f^2 is

$$\hat{f}^2 = \frac{\text{SS}_{\text{trt}}}{\text{SS}_{\text{res}}} = \frac{k-1}{N-k} \cdot F = \frac{N}{N-k} \cdot \frac{1}{k \cdot \hat{\sigma}_e^2}\sum_{i=1}^{k}\hat{\alpha}_i^2 \, ,$$

where F is the observed value of the F-statistic, yielding $\hat{f}^2 = 3.77$ for our example; the factor $N/(N-k)$ removes the bias.

The two effect sizes f^2 and η_{trt}^2 are translated into each other via

$$f^2 = \frac{\eta_{\text{trt}}^2}{1 - \eta_{\text{trt}}^2} \quad \text{and} \quad \eta_{\text{trt}}^2 = \frac{f^2}{1 + f^2} \, . \tag{4.1}$$

Much like the omnibus F-test, all three effect sizes quantify *any* pattern of group mean differences, and do not distinguish if each group deviates slightly from the grand mean, or if one group deviates by a substantial amount while the remaining do not.

4.4 Power Analysis and Sample Size for Omnibus F-test

We are often interested in determining the necessary sample size such that the omnibus F-test reaches the desired power for a given significance level. Just as before, we need four out of the following five quantities for a power analysis: the two error probabilities α and β, the residual variance σ_e^2, the sample size per group n (we only consider balanced designs), and a measure of the minimal relevant effect size (raw or standardized).

4.4.1 General Idea

Recall that under the null hypothesis of equal treatment group means, the deviations $\alpha_i = \mu_i - \mu$ are all zero, and $F = \mathrm{MS}_{\mathrm{trt}}/\mathrm{MS}_{\mathrm{res}}$ follows an F-distribution with $k - 1$ numerator and $N - k$ denominator degrees of freedom.

If the treatment effects are not zero, then the test statistic follows a *noncentral F-distribution* $F_{k-1,N-k}(\lambda)$ with noncentrality parameter

$$\lambda = n \cdot k \cdot f^2 = \frac{n \cdot k \cdot b^2}{\sigma_e^2} = \frac{n}{\sigma_e^2} \cdot \sum_{i=1}^{k} \alpha_i^2 \ .$$

The noncentrality parameter is thus the product of the overall sample size $N = n \cdot k$ and the standardized effect size f^2 (which can be translated from and to η_{trt}^2 via Eq. (4.1)). For $k = 2$, this reduces to the previous case since $f^2 = d^2/4$ and $t_n^2(\eta) = F_{1,n}(\lambda = \eta^2)$.

The idea behind the power analysis is the same as for the t-test. If the omnibus hypothesis H_0 is true, then the F-statistic follows a central F-distribution with $k - 1$ and $N - k$ degrees of freedom, shown in Fig. 4.2 (top) for two sample sizes $n = 2$ (left) and $n = 10$ (right). The hypothesis is rejected at the significance level $\alpha = 5\%$ (black shaded area) whenever the observed F-statistic is larger than the 95% quantile $F_{1-\alpha, k-1, N-k}$ (dashed line).

If H_0 is false, then the F-statistic follows a noncentral F-distribution. Two corresponding examples are shown in Fig. 4.2 (bottom) for an effect size $f^2 = 0.34$ corresponding, for example, to a difference of $\delta = 2$ between the first and the second

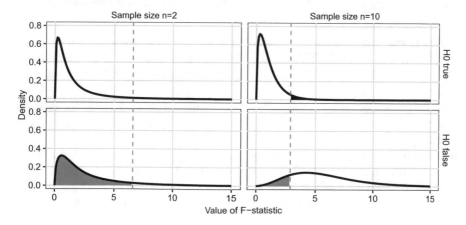

Fig. 4.2 Distribution of F-statistic if H_0 is true (top) with false positives (black shaded area), respectively, if H_0 is false and the first two groups differ by a value of two (bottom) with false negatives (gray shaded area). The dashed line indicates the 95% quantile of the central F-distribution. Left: sample size $n = 2$; right: sample size $n = 10$

treatment group, with no difference between the remaining two treatment groups. We observe that this distribution shifts to higher values with increasing sample size, since its noncentrality parameter $\lambda = n \cdot k \cdot f^2$ increases with n. For $n = 2$, we have $\lambda = 2 \cdot 4 \cdot f^2 = 2.72$ in our example, while for $n = 10$, we already have $\lambda = 10 \cdot 4 \cdot f^2 = 13.6$. In each case, the probability β of falsely not rejecting H_0 (a false positive) is the gray shaded area under the density up to the rejection quantile $F_{1-\alpha, k-1, N-k}$ of the central F-distribution. For $n = 2$, the corresponding power is then $1 - \beta = 13\%$ which increases to $1 - \beta = 85\%$ for $n = 10$.

4.4.2   Defining the Minimal Effect Size

The number of treatments k is usually predetermined, and we then exploit the relation between noncentrality parameter, effect size, and sample size for the power analysis. A frequent challenge in practice concerns defining a reasonable minimal effect size f_0^2 or b_0^2 that we want to reliably detect. Using a minimal raw effect size also requires an estimate of the residual variance σ_e^2 from previous data or a dedicated preliminary experiment.

A simple method to provide a minimal effect size uses the fact that $f^2 \geq d^2/2k$ for the standardized effect size d between any pair of group means. The standardized difference $d_0 = \delta_0/\sigma = (\mu_{\max} - \mu_{\min})/\sigma$ between the largest and smallest group means therefore provides a conservative minimal effect size $f_0^2 = d_0^2/2k$ (Kastenbaum et al. 1970).

We can improve on the inequality for specific cases, and Cohen proposed three patterns with minimal, medium, and maximal variability of treatment group differences α_i, and provided their relation to the minimal standardized difference d_0 (Cohen 1988, p. 276ff).

- If only two groups show a deviation from the common mean, we have $\alpha_{\max} = +\delta_0/2$ and $\alpha_{\min} = -\delta_0/2$ for these two groups, respectively, while $\alpha_i = 0$ for the $k - 2$ remaining groups. Then, $f_0^2 = d_0^2/2k$ and our conservative effect size is in fact exact.
- If the group means μ_i are equally spaced with distances $\delta_0/(k - 1)$, then the omnibus effect size is $f_0^2 = d_0^2/4 \cdot (k + 1)/3(k - 1)$. For $k = 4$ and $d = 3$, an example is $\mu_1 = 1$, $\mu_2 = 2$, $\mu_3 = 3$, and $\mu_4 = 4$.
- If half of the groups is at one extreme $\alpha_{\max} = +\delta_0/2$ while the other half is at the other extreme $\alpha_{\min} = -\delta_0/2$, then $f_0^2 = d_0^2/4$ if k is even and $f_0^2 = d_0^2/4 \cdot (1 - 1/k^2)$ if k is odd. Again for $k = 4$ and $d = 3$, an example is $\alpha_1 = \alpha_3 = -1.5$, $\alpha_2 = \alpha_4 = +1.5$. For $\mu = 10$ and $\sigma^2 = 1$, this corresponds to $\mu_1 = \mu_3 = 8.5$ and $\mu_2 = \mu_4 = 11.5$.

4.4.3 Calculating Power

A simple power analysis function for R is given in Sect. 4.7 for illustration, while the more flexible built-in procedure `power.anova.test()` directly provides the necessary calculations. This procedure accepts the number of groups k (`groups=`), the per-group sample size n (`n=`), the residual variance σ_e^2 (`within.var=`), the power $1 - \beta$ (`power=`), the significance level α (`sig.level=`), and a modified version of the raw effect size $\nu^2 = \sum_i \alpha_i^2/(k - 1)$ (`between.var=`) as its arguments. Given any four of these parameters, it will calculate the remaining one.

We look at a range of examples to illustrate the use of power analysis in R. We assume that our previous analysis was completed and we intend to explore new experiments of the same type. This allows us to use the variance estimate $\hat{\sigma}_e^2$ as our assumed within-group variance for the power analyses, where we round this estimate to $\hat{\sigma}_e^2 = 1.5$ for the following calculations. In all examples, we set our false positive probability to the customary $\alpha = 5\%$.

Raw Effect Size b^2

From a minimal raw effect size b_0^2, we find the corresponding `between.var` argument as

$$\nu_0^2 = \frac{k}{k - 1} \cdot b_0^2 .$$

For example, we might consider an effect with $\alpha_1 = \alpha_2 = +1$ and $\alpha_3 = \alpha_2 = -1$, such that $D1$ and $D2$ yield identically higher average enzyme levels, while $D3$ and $D4$ yield correspondingly lower enzyme levels. Then our raw effect size is $b_0^2 = ((+1)^2 + (+1)^2 + (-1)^2 + (-1)^2)/4 = 1$ and we use `between.var= 1.33` and `within.var= 1.5` for our calculation. For a required power of 80%, this yields $n = 5$ mice per group. Using only $n = 2$ mice per group, we achieve a power of $1 - \beta = 21\%$.

Standardized Effect Size f^2

From a minimal standardized effect size f_0^2, we find the corresponding `between.var` argument as

$$\nu_0^2 = \frac{k}{k - 1} \cdot \sigma_e^2 \cdot f_0^2 ,$$

but this formula defeats the purpose of standardized effect size, since it explicitly requires the residual variance. The solution to this problem comes from noticing that since we measure effects in units of standard deviation, we can set $\sigma_e^2 = 1$ in this formula and use it in conjunction with `within.var=1` to achieve the desired result.

The standardized effect sizes of $f_0^2 = 0.01$ (small), $f_0^2 = 0.06$ (medium), and $f_0^2 = 0.16$ (large) then translate to $\nu_0^2 = 0.01$, $\nu_0^2 = 0.08$, and $\nu_0^2 = 0.21$, respectively.

For a sample size of 10 mice per drug, these minimal effect sizes correspond to a power of 7%, 21%, and 50%, respectively, and to achieve a desired power of 80% would require 274, 45, and 18 mice per drug, respectively.

Range of Group Mean Differences

For finding the required sample size based on a minimal group difference, we plan a follow-up experiment to detect a difference of at least $\delta_0 = 1$ between the average enzyme levels of $D1$ and $D2$. With our assumed variance of $\sigma_e^2 = 1.5$, this corresponds to a standardized effect size of $d_0 = \delta_0/\sigma = 0.82$.

We use Cohen's first pattern and set $\alpha_1 = \delta_0/2, \alpha_2 = -\delta_0/2, \alpha_3 = \alpha_4 = 0$, which yields a standardized effect size of $f^2 = 0.083$, respectively, $\eta^2 = 0.077$. Using the formulas above, this corresponds to a `between.var` parameter of $\nu_0^2 = 0.17$.

The required sample size for a power of 80% is 34 mice per group.

Standardized Effect Size f^2

To plan an experiment that yields $\eta_{0,\text{drug}}^2 = 20\%$ of the variation explained by the differences in drugs, we proceed as for f_0^2 and translate this effect size into the `between.var` argument by

$$\nu_0^2 = \frac{k}{k-1} \cdot \frac{\eta_{0,\text{drug}}^2}{1 - \eta_{0,\text{drug}}^2} \, ,$$

using $\sigma_e^2 = 1$ here and as our `within.var` argument.

For $k = 4$ groups, this yields a parameter of $\nu_0^2 = 0.13$, and we can detect this effect size with power $1 - \beta = 44\%$ for a sample size of $n = 10$ per treatment group. Similarly, we require $n = 20$ mice per drug treatment to achieve a desired power of 80%.

Minimal Detectable Effect Size

We can also find the minimal effect size detectable for a given sample size and power. For example, we might plan a similar experiment with the same four drugs, but we have only maximally 20 mice per drug available. Plugging everything in, we find a detectable between-groups variation of $\nu_0^2 = 0.29$, corresponding to an effect size of $f^2 = 0.14$ or $\eta_{drug}^2 = 13\%$.

4.4.4 Power Analysis in Practice

The power calculations for the omnibus F-test rely on the same assumptions as the test itself, and require identical residual variances for each treatment group and normally distributed residuals. Results are only marginally affected by moderately different group variances or moderate deviations from normality, but the seemingly precise six decimals of the built-in `power.anova.test()` should not be taken too literally. More severe errors result if observations are not independent, for example, if correlations arise by measuring the same unit multiple times.

Even if all assumptions are matched perfectly, the calculated sample size is still based on an educated guess or a previous estimate of the residual variance. We should therefore make an allowance for a margin of error in our calculations.

We can again use a conservative approach and base our power calculations on the upper confidence limit rather than the variance estimate itself. For our example, the residual variance is $\hat{\sigma}_e^2 \approx 1.5$ with a 95%-confidence interval of [0.94, 2.74]. For $k = 4$ groups, a desired power of 80% and a significance level of 5%, detecting a raw effect of $b_0^2 = 0.12$ ($f_0^2 = 0.08$), requires 34 mice per group. Taking the upper confidence limit, the required sample size increases by a factor of roughly UCL/$\hat{\sigma}_e^2 \approx$ $2.74/1.5 = 1.8$ to 61 mice per group.

A less-conservative approach increases the 'exact' sample size by 20–30%; for our example, this yields a sample size of 40 for a 20% margin, and of 44 for a 30% margin, compared to the original exact sample size of 34.

4.4.5 Portable Power

The *portable power* procedure exploits the fact that for the common significance level $\alpha = 5\%$ and a commonly desired power of $1 - \beta = 80\%$, the noncentrality parameter λ changes comparatively little, allowing us to use a crude approximation for our calculations (Wheeler 1974). Such a procedure is very helpful in the early stages of planning an experiment, when all that is needed are reasonably accurate approximations for sample sizes to gauge the practical implications of an experiment design.

The portable power procedure uses the quantity

$$\phi^2 = \lambda/k = n \cdot f^2 .$$

This quantity is reasonably well approximated by $\phi^2 = 5$ if we expect a few (less than 10, say) denominator degrees of freedom and by $\phi^2 = 3$ if we expect many such degrees of freedom. It becomes smaller with an increasing number of treatment groups, but this dependency is often negligible in practice. A reasonable strategy is to calculate the sample size $n = \phi^2/f^2$ assuming $\phi^2 = 3$. If the resulting n is small, we should repeat the calculation with $\phi^2 = 5$ and use the resulting larger sample size.

We illustrate the procedure by revisiting two previous examples. Recall that for $k = 4$ and a minimal difference of $\delta_0 = 1$, we found an effect size of $f^2 = 0.083$. The exact power analysis for $\alpha = 5\%$ and $1 - \beta = 80\%$ indicated a required sample size of 34. Assuming that the required sample size is sufficiently large, we approximate this analysis using $\phi^2 = 3$, which yields a sample size of $n = \phi^2/f^2 = 3/0.083 \approx 36$ and overestimates the exact value by about 7%.

Using again $\alpha = 5\%$ and $1 - \beta = 80\%$, a sample size of $n = 20$ mice in each of the four treatment groups showed a minimal detectable effect size of $f^2 = 0.14$ using the exact calculation. Using the approximation $\phi^2 = 3$, the portable power calculation yields $f^2 = \phi^2/n = 0.15$, an error of about 4%.

The portable power approximations are helpful for getting an idea if a necessary minimal effect size is within reach given our resources, and for finding a rough estimate of the minimal effect detectable given a specific experiment size. The approximation error is typically less than 35% and often considerably lower. Given that the variance estimate is often much less precise, this magnitude of error should be acceptable for a crude calculation in most circumstances. Once a suitable design and sample size are identified, we should of course use the exact methods and an error margin to determine the final sample size.

4.5   Hasse Diagrams and Linear Model Specification

The analysis of variance has an intimate connection with classical linear regression and both methods are based on describing the observed data by a linear mathematical model (cf. Sect. 4.7). The analysis of more complex designs becomes relatively straightforward when this connection is exploited, and most statistical software will internally run a linear regression procedure for computing an ANOVA. While this relieves the practitioner from much tedious algebra, it still means that the appropriate linear model for an experimental design has to be correctly specified for the software.

The specification has two parts: first, the experimental design has to be translated into the linear model, such that the statistical inferences fully capture the logical structure of our experiment. And second, the linear model has to be translated into a model specification in the software. We can solve both problems by *Hasse diagrams* that visualize the logical structure of an experiment, and from which both a linear model formula and a symbolic representation of the model can be derived with relative ease. We already saw some simple examples of these diagrams in Figs. 2.4, 2.7 and 2.8. We now work out the connection between design, diagram, and model more systematically. Some of the following discussions might seem overly complicated for the relatively simple designs discussed so far, but are necessary for more complex designs in the following chapters.

4.5.1 Hasse Diagrams of Experiment Structure

We introduce some additional terminology to be able to precisely describe the logical structure of an experiment.

Treatment, Unit, and Response Structures

The *treatment structure* of an experiment describes the *treatment factors* and their relationships. In our drug example, the experiment has a single treatment factor **Drug** with four *levels* $D1$, $D2$, $D3$, and $D4$. Other designs use several treatment factors, and each applied treatment is then a combination of one level from each treatment factor.

The *unit (or design) structure* describes the *unit factors* and their relationships. A *unit factor* logically organizes the experimental material, and our experiment has a single unit factor *(Mouse)* with 32 levels, each level corresponding to one mouse. Unit factors are of several basic types: the smallest subdivision of the experimental material to which levels of a treatment factor are randomly assigned is called the *experimental unit* of this treatment factor; it provides the residual variance for testing this treatment factor.

Groups of units are specified by a *grouping factor*, also known as a *blocking factor*; these are often *non-specific* and of no direct inferential interest, but are used to remove variation from comparisons or take account of units in the same group being more similar than units in different groups. A blocking factor can also be *intrinsic* and describe a non-randomizable property of another unit factor; a common example is the sex of an animal, which we cannot deliberately choose (so it is not a treatment), but which we need to keep track of in our inferences.

Treatment factors are often *fixed factors* with predetermined fixed levels, while unit factors are often *random factors* whose levels are a random sample from a population; in a replication of the experiment, the fixed factor levels would remain the same (we use the same four drugs again), while the random factor levels change (we do not use the same mice again). We denote treatment factors by an informative name written in **bold** and unit factors in *italics*; we denote random factors by parentheses: the treatment factor **Drug** is fixed for our experiment, while the unit factor *(Mouse)* is random.

The observations are recorded on the *response unit factor*, and we mainly consider experiments with a simple *response structure* where a single value is observed on one unit factor in the design, which we denote by underlining.

The treatment and unit structures are created by *nesting* and *crossing* of factors. A factor A is *crossed* with another factor B if each level of A occurs together with each

A

(Mouse)	Vendor	
	A	B
1	$y_{A,1}$	$y_{B,1}$
2	$y_{A,2}$	$y_{B,2}$
3	$y_{A,3}$	$y_{B,3}$

B

(Mouse)	Drug			
	D1	D2	D3	D4
1			$y_{D3,1}$	
2				$y_{D4,2}$
3		$y_{D2,1}$		
4	$y_{D1,1}$			
5			$y_{D3,2}$	
6		$y_{D2,2}$		
7				$y_{D4,1}$
8	$y_{D1,2}$			

Fig. 4.3 Examples of data layouts. **A** Factors Vendor and Mouse are crossed. **B** Factor Mouse is nested in factor Drug

level of *B* and vice versa. This implicitly defines a third *interaction factor* denoted by *A:B*, whose levels are the possible combinations of levels of *A* with levels of *B*. In our paired design (Fig. 2.8), the treatment factor **Vendor** is crossed with *(Mouse)*, since each kit (that is, each level of **Vendor**) is assigned to each mouse. We omitted the interaction factor, since it coincides with *(Sample)* in this case. The data layout for two crossed factors is shown in Fig. 4.3A; the cross-tabulation is completely filled.

A factor *B* is *nested* in another factor *A* if each level of *B* occurs together with one and only one level of *A*. For our current example, the factor *(Mouse)* is nested in **Drug**, since we have one or more mice per drug, but each mouse is associated with exactly one drug; Fig. 4.3B illustrates the nesting for two mice per drug.

When designing an experiment, the treatment structure is determined by the purpose of the experiment: what experimental factors to consider and how to combine factor levels to treatments. The unit structure is then used to accommodate the treatment structure and to maximize precision and power for the intended comparisons. In other words:

> **The treatment structure is the driver in planning experiments, the design structure is the vehicle.** (van Belle 2008, p180)

Finally, the *experiment structure* combines the treatment and unit structures and is constructed by making the randomization of each treatment on its experimental unit explicit. It provides the logical arrangement of units and treatments.

Constructing the Hasse Diagrams

To emphasize the two components of each experimental design, we draw separate Hasse diagrams for the treatment and unit structures, which we then combine into

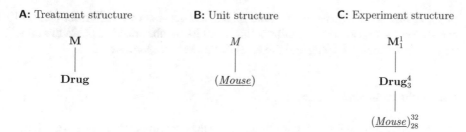

A: Treatment structure   **B:** Unit structure   **C:** Experiment structure

M   *M* M_1^1

Drug   *(Mouse)* $\mathbf{Drug}_3^4$

$(Mouse)_{28}^{32}$

Fig. 4.4 Hasse diagrams for a completely randomized design for determining the effect of four different drugs using 8 mice per drug and a single response measurement per mouse

the experiment structure diagram by considering the randomization. The Hasse diagram visualizes the nesting/crossing relations between the factors. Each factor is represented by a node, shown as the factor name. If factor B is nested in factor A, we write B below A and connect the two nodes with an edge. The diagram is thus 'read' from top to bottom. If A and B are crossed, we write them next to each other and connect each to the next factor that it is nested in. We then create a new factor denoted by $A:B$, whose levels are the combinations of levels of A with levels of B, and draw one edge from A and one edge from B to this factor. Each diagram has a top node called $\mathbf{M}$ or M, which represents the grand mean, and all other factors are nested in this top node.

The Hasse diagrams for our drug example are shown in Fig. 4.4. The treatment structure contains the single treatment factor **Drug**, nested in the obligatory top node $\mathbf{M}$ (Fig. 4.4A). Similarly, the unit structure contains only the factor *(Mouse)* nested in the obligatory top node M (Fig. 4.4B).

We construct the experiment structure diagram as follows: first, we merge the two top nodes $\mathbf{M}$ and M of the treatment and unit structure diagram, respectively, into a single node $\mathbf{M}$. We then draw an edge from each treatment factor to its experimental unit factor. If necessary, we clean up the resulting diagram by removing unnecessary 'shortcut' edges: whenever there is a path A–B–C, we remove the edge A–C if it exists since its nesting relation is already implied by the path.

In our example, we merge the two nodes $\mathbf{M}$ and M into a single node $\mathbf{M}$. Both **Drug** and *(Mouse)* are now nested under the same top node. Since **Drug** is randomized on *(Mouse)*, we write *(Mouse)* below **Drug** and connect the two nodes with an edge. This makes the edge from $\mathbf{M}$ to *(Mouse)* redundant and we remove it from the diagram (Fig. 4.4C).

We complete the diagram by adding the number of levels for each factor as a superscript to its node, and by adding the degrees of freedom for this factor as a subscript. The degrees of freedom for a factor A are calculated as the number of levels minus the degrees of freedom of each factor that A is nested in. The number of levels and the degrees of freedom of the top node $\mathbf{M}$ are both one.

The superscripts are the number of factor levels: 1 for $\mathbf{M}$, 4 for **Drug**, and 32 for *(Mouse)*. The degrees of freedom for **Drug** are therefore $3 = 4 - 1$, the number of levels minus the degrees of freedom of **M**. The degrees of freedom for *(Mouse)* are 28, which we calculate by subtracting from its number of levels (32) the three degrees of freedom of **Drug** and the single degree of freedom for **M**.

Hasse Diagrams and F-Tests

We can derive the omnibus F-test directly from the experiment diagram in Fig. 4.4C: the omnibus null hypothesis claims equality of the group means for the treatment factor **Drug**. This factor has four such means (given by the superscript), and the F-statistic has three numerator degrees of freedom (given by the subscript). For any experimental design, we find the corresponding experimental unit factor that provides the within-group variance σ_e^2 in the diagram by starting from the factor tested and moving downwards along edges until we find the first random factor. In our example, this trivially leads to identifying *(Mouse)* as the relevant factor, providing $N - k = 28$ degrees of freedom (the subscript) for the F-denominator. The test thus compares MS_{drug} to MS_{mouse}.

An Example with Sub-sampling

In biological experimentation, experimental units are frequently sub-sampled, and the data contain several response values for each experimental unit. In our example, we might still randomize the drug treatments on the mice, but take four blood samples instead of one from each mouse and measure them independently. Then, the mice are still the experimental units for the treatment, but the blood samples now provide the response units. The Hasse diagrams in Fig. 4.5 illustrate this design.

The treatment structure is identical to our previous example, and contains **Drug** as its only relevant factor. The unit structure now contains a new factor *(Sample)* with 128 levels, one for each measured enzyme level. It is the response factor that

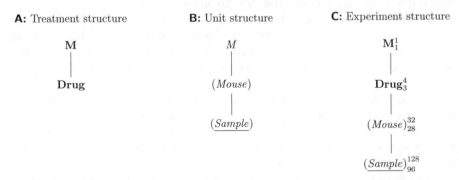

A: Treatment structure   **B:** Unit structure   **C:** Experiment structure

Fig. 4.5 Completely randomized design for determining the effect of four different drugs using 8 mice per drug and four samples measured per mouse

provides the observations. Since each sample belongs to one mouse, and each mouse has several samples, the factor *(Sample)* is nested in *(Mouse)*. The observations are then partitioned first into 32 groups—one per mouse—and further into 128—one per sample per mouse. For the experiment structure, we randomize **Drug** on *(Mouse)*, and arrive at the diagram in Fig. 4.5C.

The F-test for the drug effect again uses the mean squares for **Drug** on 3 degrees of freedom. Using our rule, we find that *(Mouse)*—and not *(Sample)*—is the experimental unit factor that provides the estimate of the variance for the F-denominator on 28 degrees of freedom. As far as this test is concerned, the 128 samples are technical replicates or pseudo-replicates. They do not reflect the biological variation against which we need to test the differences in enzyme levels for the four drugs, since drugs are randomized on mice and not on samples.

4.5.2 The Linear Model

For a completely randomized design with k treatment groups, we can write each datum y_{ij} explicitly as the corresponding treatment group mean and a random deviation from this mean:

$$y_{ij} = \mu_i + e_{ij} = \mu + \alpha_i + e_{ij} . \tag{4.2}$$

The first model is called a *cell means model*, while the second, equivalent, model is a *parametric model*. If the treatments had no effect, then all $\alpha_i = \mu_i - \mu$ are zero and the data are fully described by the grand mean μ and the residuals e_{ij}. Thus, the parameters α_i measure the systematic difference of each treatment from the grand mean and are independent of the experimental units.

It is crucial for an analysis that the linear model fully reflects the structure of the experiment. The Hasse diagrams allow us to derive an appropriate model for any experimental design with comparative ease. For our example, the diagram in Fig. 4.4C has three factors: **M**, **Drug**, and *(Mouse)*, and these are reflected in the three sets of parameters μ, α_i, and e_{ij}. Note that there are four parameters α_i to produce the four group means, but given three and the grand mean μ, the fourth parameter can be calculated; thus, there are four parameters α_i, but only three can be independently estimated given μ, as reflected by the three degrees of freedom for **Drug**. Further, the e_{ij} are 32 random variables, and this is reflected in the fact that *(Mouse)* is a random factor. Given estimates for μ and α_i, the e_{ij} in each of the four groups must sum to zero and only 28 values are independent.

For the sub-sampling example in Fig. 4.5, the linear model is

$$y_{ijk} = \mu + \alpha_i + m_{ij} + e_{ijk} ,$$

where m_{ij} is the average deviation of measurements of mouse j in treatment group i from the treatment group mean, and e_{ijk} are the deviations of individual measurements of a mouse to its average. These terms correspond exactly to **M**, **Drug**, *(Mouse)*, and *(Sample)*.

4.5.3 Analysis of Variance in R

The aov() function provides all the necessary functionality for calculating complex ANOVAs and for estimating the model parameters of the corresponding linear models. It requires two arguments: data= indicates a data-frame with one column for each variable in the model, and aov() uses the values in these columns as the input data. The model is specified with the formula= argument using a *formula*. This formula describes the factors in the model and their crossing and nesting relationships, and can be derived directly from the experiment diagram.

For our first example, the data is stored in a data-frame called drugs which consists of 32 rows and three columns: y contains the observed enzyme level, drug the drug (D1 to D4), and mouse the number of the corresponding mouse ($1 \ldots 32$). Our data are then analyzed using the command aov(formula=y~1+drug+Error (mouse), data=drugs).

The corresponding formula has three parts: on the left-hand side, the name of the column containing the observed response values (y), followed by a tilde. Then, a part describing the fixed factors, which we can usually derive from the treatment structure diagram: here, it contains the special symbol 1 for the grand mean μ and the term drug encoding the four parameters α_i. Finally, the Error() part describes the random factors and is usually equivalent to the unit structure of the experiment. Here, it contains only mouse. An R formula can often be further simplified; in particular, aov() will always assume a grand mean 1, unless it is explicitly removed from the model, and will always assume that each row is one observation relating to the lowest random factor in the diagram. Both parts can be skipped from the formula and are implicitly added; our formula is thus equivalent to y~drug. We can read a formula as an instruction: explain the observations y_{ij} in column y by the factors on the right: a grand mean $1/\mu$, the refinements drug/α_i that give the group means when added to the grand mean, and the residuals mouse/e_{ij} that cover each difference between group mean and actual observation.

The function aov() returns a complex data structure containing the fitted model. Using summary(), we produce a human-readable ANOVA table:

```
m = aov(y~drug, data=drugs)
summary(m)
```

	Df	Sum Sq	Mean Sq	F value	Pr(>F)
Drug	3	155.89	51.96	35.17	1.24e-09
Residuals	28	41.37	1.48		

It corresponds to our manually computed Table 4.2. Each factor in the diagram produces one row in the ANOVA table, with the exception of the trivial factor **M**. Moreover, the degrees of freedom correspond between each diagram factor and its table row. This provides a quick and easy check if the model formula correctly describes the experiment structure. While not strictly necessary here, this is an important and useful feature of Hasse diagrams for more complex experimental designs.

From Hasse Diagram to Model Specification in R

From the Hasse diagrams, we construct the formula for the ANOVA as follows:

- There is one variable for each factor in the diagram;
- terms are added using +;
- R adds the factor **M** implicitly; we can make it explicit by adding 1;
- if factors A and B are crossed, we write A*B or equivalently A + B + A:B;
- if factor B is nested in A, we write A/B or equivalently A + A:B;
- the formula has two parts: one for the random factors inside the Error() term, and one for the fixed factors outside the Error() term;
- in most cases, the unit structure describes the random factors and we can use its diagram to derive the Error() formula;
- likewise, the treatment structure usually describes the fixed factors, and we can use its diagram to derive the remaining formula.

In our sub-sampling example (Fig. 4.5), the unit structure contains *(Sample)* nested in *(Mouse)*, and we describe this nesting by the formula mouse/sample. The formula for the model is thus y~1+drug+Error(mouse/sample), respectively, y~drug+Error(mouse).

The aov() function does not directly provide estimates of the group means, and an elegant way of estimating them in R is the emmeans() function from the emmeans package.[1] It calculates the *expected marginal means* (sometimes confusingly called *least squares means*) as defined in (Searle and Milliken 1980) for any combination of experiment factors using the model found by aov(). In our case, we can request the estimated group averages $\hat{\mu}_i$ for each level of drug, which emmeans() calculates from the model as $\hat{\mu}_i = \hat{\mu} + \hat{\alpha}_i$:

```
em = emmeans(m, ~drug)
```

This yields the estimates, their standard errors, degrees of freedom, and 95%-confidence intervals in Table 4.3.

---

[1]This package was previously called lsmeans (Lenth 2016).

Table 4.3 Estimated cell means, standard errors, and 95%-confidence intervals

	Mean	se	df	LCL	UCL
D1	14.33	0.43	28	13.45	15.21
D2	12.81	0.43	28	11.93	13.69
D3	9.24	0.43	28	8.36	10.12
D4	9.34	0.43	28	8.46	10.22

4.6 Unbalanced Data

When designing an experiment with a single treatment factor, it is rather natural to consider using the same number of experimental units for each treatment group. We saw in Sect. 3.2 that this also leads to the lowest possible standard error when estimating the difference between any two treatment means, provided that the ANOVA assumption of equal variance in each treatment group holds.

Sometimes, such a *fully balanced* design is not achieved, because the number of available experimental units is not a multiple of the number of treatment groups. For example, we might consider studying the four drugs again, but with only 30 mice at our disposal rather than 32. Then, we either need to reduce each treatment group to seven mice, leaving two mice, or use eight mice in two treatment groups, and seven in the other two (alternatively eight mice in three groups and six in the remaining).

Another cause for missing full balance is that some experimental units fail to give usable response values during the experiment or their recordings go missing. This might happen because some mice die (or escape) during the experiment, a sample gets destroyed or goes missing, or some faulty readings are discovered after the experiment is finished.

In either case, the number of responses per treatment group is no longer the same number n, but each treatment group has its own number of experimental units n_i, with a total experiment size $N = n_1 + \cdots + n_k$ for k treatment groups.

4.6.1 Analysis of Variance

The one-way analysis of variance is little influenced by unbalanced data, provided each treatment group has at least one observation. The total sum of squares again decomposes into a treatment and a residual term, both of which now depend on n_i in the obvious way:

$$
\text{SS}_{\text{tot}} = \sum_{i=1}^{k} \sum_{j=1}^{n_i} (y_{ij} - \bar{y}_i + \bar{y}_i - \bar{y})^2 = \underbrace{\sum_{i=1}^{k} n_i \cdot (\bar{y}_i - \bar{y})^2}_{\text{SS}_{\text{trt}}} + \underbrace{\sum_{i=1}^{k} \sum_{j=1}^{n_i} (y_{ij} - \bar{y}_i)^2}_{\text{SS}_{\text{res}}} .
$$

Clearly, if no responses are observed for a treatment group, its average response cannot be estimated, and no inference is possible about this group or its relation to other groups.

The ratio of treatment to residual mean squares then again forms the F-statistic

$$F = \frac{SS_{trt}/(k-1)}{SS_{res}/(N-k)}$$

which has an F-distribution with $k-1$ and $N-k$ degrees of freedom under the null hypothesis $H_0 : \mu_1 = \cdots = \mu_k$ that all treatment group means are equal.

4.6.2 Estimating the Grand Mean

With unequal numbers n_i of observations per cell, we now have two reasonable estimators for the grand mean μ: one estimator is the *weighted mean* $\bar{y}_{..}$, which is the direct equivalent to our previous estimator:

$$\bar{y}_{..} = \frac{y_{..}}{N} = \frac{1}{N}\sum_{i=1}^{k}\sum_{j=1}^{n_i} y_{ij} = \frac{n_1 \cdot \bar{y}_{1.} + \cdots n_k \cdot \bar{y}_{k.}}{n_1 + \cdots + n_k} = \frac{n_1}{N}\cdot\bar{y}_{1.} + \cdots + \frac{n_k}{N}\cdot\bar{y}_{k.} .$$

This estimator weighs each estimated treatment group mean by the number of available response values and hence its value depends on the number of observations per group via the weights n_i/N. Its variance is

$$\mathrm{Var}(\bar{y}_{..}) = \frac{\sigma^2}{N} .$$

The weighted mean is often an undesirable estimator, because larger groups then contribute more to the estimation of the grand mean. In contrast, the *unweighted mean*

$$\tilde{y}_{..} = \frac{1}{k}\sum_{i=1}^{k}\left(\frac{1}{n_i}\sum_{j=1}^{n_i} y_{ij}\right) = \frac{\bar{y}_{1.} + \cdots + \bar{y}_{k.}}{k}$$

first calculates the average of each treatment group based on the available observations, and then takes the mean of these group averages as the grand mean. This is precisely the estimated marginal mean, an estimator for the *population marginal mean* $\mu = (\mu_1 + \cdots + \mu_k)/k$.

In the direct extension to our discussion in Sect. 3.2, its variance is

$$\mathrm{Var}(\tilde{y}_{..}) = \frac{\sigma^2}{k^2}\cdot\left(\frac{1}{n_1} + \cdots + \frac{1}{n_k}\right) ,$$

which is minimal if $n_1 = \cdots = n_k$ and then reduces to the familiar σ^2/N.

These two estimators yield the same result if the data are balanced and $n_1 = \cdots = n_k$, but their estimates differ for unbalanced data. For example, taking the first 4, 2, 1, and 2 responses for $D1$, $D2$, $D3$, and $D4$ from Table 4.1 yields a fairly unbalanced design in which $D3$ (with comparatively low responses) is very underrepresented compared to $D1$ (with high value), for example. The two estimators are $\bar{y}_{..} = 12.5$ and $\tilde{y}_{..} = 11.84$. The average on the full data is 11.43.

4.6.3   Degrees of Freedom

With different numbers of observations between groups, the resulting degrees of freedom with k treatment groups and a total of N observations are still $k - 1$ for the treatment factor, and $N - k$ for the residuals for designs without sub-sampling. If sub-sampling of experimental units is part of the design, however, the degrees of freedom in the F-test statistic can no longer be determined exactly. A common approximation is the conceptually simple *Satterthwaite approximation* based on a weighted mean (Satterthwaite 1946), while the more conservative *Kenward-Roger approximation* (Kenward and Roger 1997) provides degrees of freedom that agree with the Hasse diagram for more complex models that we discuss later.

4.7   Notes and Summary

Notes

Standard analysis of variance requires normally distributed, independent data in each treatment group and homogeneous group variances. Moderate deviations from normality and moderately unequal variances have little impact on the F-statistics, but non-independence can have devastating effects (Cochran 1947; Lindman 1992). The method by Kruskal and Wallis provides an alternative to the one-way ANOVA based on ranks of observations rather than an assumption of normally distributed data (Kruskal and Wallis 1952).

Effect sizes for ANOVA are comparatively common in psychology and related fields (Cohen 1988; Lakens 2013), but have also been advertised for biological sciences, where they are much less used (Nakagawa and Cuthill 2007). Like any estimate, standardized effect sizes should also be reported with a confidence interval, but standard software rarely provides out-of-the-box calculations (Kelley 2007); these confidence intervals are calculated based on noncentral distributions, similar to our calculations in Sect. 2.5 for Cohen's d (Venables 1975; Cumming and Finch 2001). The use of its upper confidence limit rather than the variance's point estimate for power analysis is evaluated in Kieser and Wassmer (1996). A pilot study can be conducted to estimate the variance and enable informed power calculations for the main experiment; a group size of $n = 12$ is recommended for clinical pilot trials (Julious

2005). The use of portable power analysis makes use of ϕ^2 for which tables were originally given in Tang (1938).

A Simple Function for Calculating Sample Size

The following code illustrates the power calculations from Sect. 4.4. It accepts a variety of effect size measures (b^2, f^2, η^2, or λ), either the numerator degrees of freedom `df1=` or the number of treatment groups `k=`, the denominator degrees of freedom `df2=` or the number of samples `n=`, and the significance level `alpha=` and calculates the power of the omnibus F-test. This code is meant to be more instructional than practical, but its more versatile arguments will also allow us to perform power analysis for linear contrasts, which is not possible with the built-in `power.anova.test()`. For a sample size calculation, we would start with a small n, use this function to find the power, and then increment n until the desired power is exceeded.

```
# Provide EITHER b2 (with s) OR f2 OR lambda OR eta2
# Provide EITHER k OR df1,df2
getPowerF = function(n, k=NULL, df1=NULL, df2=NULL,
                     b2=NULL, s=NULL,
                     f2=NULL, lambda=NULL, eta2=NULL,
                     alpha) {
  # If necessary, calculate degrees of freedom
  if(is.null(df1)) df1 = k - 1
  if(is.null(df2)) df2 = n*k - k
  # Map effect size to ncp lambda=f2*n*k if necessary
  if(!is.null(b2)) lambda = b2 * n * (df1+1) / s
  if(!is.null(f2)) lambda = f2 * n * (df1+1)
  if(!is.null(eta2)) lambda = (eta2 / (1-eta2)) * n * (df1+1)
  # Critical quantile for rejection under H0
  q = qf(p=1-alpha, df1=df1, df2=df2)
  # Power
  pwr = 1 - pf(q=q, ncp=lambda, df1=df1, df2=df2)
  return(pwr)
}
```

ANOVA and Linear Regression

The one-way ANOVA is formally equivalent to a linear regression model with regressors $x_{1j}, \ldots, x_{kj}$ to encode the k group memberships:

$$y_{ij} = \mu + \alpha_i + e_{ij} = \mu + \alpha_1 \cdot x_{1j} + \cdots + \alpha_k \cdot x_{kj} + e_{ij} ,$$

and this model is used internally by R. Since it describes k group means $\mu_1, \ldots, \mu_k$ by $k + 1$ parameters $\mu, \alpha_1, \ldots, \alpha_k$, it is *overparameterized* and an additional constraint is required for finding suitable estimates (Nelder 1994, 1998). Two common constraints are the *treatment coding* and the *sum coding* which both yield (dependent) parameter estimates, but these estimates will differ in value and in interpretation. For this reason, we only use the linear model fit to calculate the estimated marginal means, such as group averages, which are independent of the coding for the parameters used in the model estimation procedure.

The *treatment coding* uses $\hat{\alpha}_r = 0$ for a specific *reference group* r; $\hat{\mu}$ then estimates the group mean of the reference group r, and $\hat{\alpha}_i$ estimates the difference between group i and the reference group r. The vector of regressors for the response y_{ij} is

$$(x_{1j}, \ldots, x_{kj}) = (0, \ldots, 0, \underbrace{1}_{i}, 0, \ldots, 0) \text{ for } i \neq r$$

$$(x_{1j}, \ldots, x_{kj}) = (0, \ldots, 0, \underbrace{0}_{i}, 0, \ldots, 0) \text{ for } i = r .$$

In R, the treatment groups are usually encoded as a factor variable, and `aov()` will use its alphabetically first level as the reference group by default. We can set the reference level manually using `relevel()` in base-R or the more convenient `fct_relevel()` from the `forcats` package. We can use the `dummy.coef()` function to extract all parameter estimates from the fitted model, and `summary.lm()` to see the estimated parameters together with their standard errors and t-tests. The treatment coding can also be set manually using the `contrasts=` argument of `aov()` together with `contr.treatment(k)`, where `k` is the number of treatment levels.

For our example, `aov()` uses the treatment level $D1$ as the reference level. This gives the estimates `(Intercept)` = $\hat{\mu} = 14.33$ for the average enzyme level in the reference group, which serves as the grand mean, $D1 = \hat{\alpha}_1 = 0$ as expected, and the three differences from $D2$, $D3$, and $D4$ to $D1$ as $D2 = \hat{\alpha}_2 = -1.51$, $D3 = \hat{\alpha}_3 = -5.09$, and $D4 = \hat{\alpha}_4 = -4.99$, respectively. We find the group mean for $D3$, for example, as $\hat{\mu}_3 = \hat{\mu} + \hat{\alpha}_3 = 14.33 + (-5.09) = 9.24$.

The *sum coding* uses the constraint $\sum_{i=1}^{k} \hat{\alpha}_i = 0$. Then, $\hat{\mu}$ is the estimated grand mean, and $\hat{\alpha}_i$ is the estimated deviation of the ith group mean from this grand mean. The vector of regressors for the response y_{ij} is

$$(x_{1j}, \ldots, x_{kj}) = (0, \ldots, 0, \underbrace{1}_{i}, 0, \ldots, 0) \text{ for } i = 1, \ldots, k-1$$

$$(x_{1j}, \ldots, x_{kj}) = (-1, \cdots, -1, 0) \text{ for } i = k .$$

The `aov()` function uses the argument `contrasts=` to specify the desired coding for each factor, and `contrasts=list(drug=contr.sum(4))` provides the sum encoding for our example. The resulting parameter estimates are `(Intercept)` = $\hat{\mu} = 11.43$ for the average enzyme level over all groups, and $D1 = \hat{\alpha}_1 = 2.9$, $D2 = \hat{\alpha}_2 = 1.38$, $D3 = \hat{\alpha}_3 = -2.19$, and $D4 = \hat{\alpha}_4 = -2.09$, respectively, as the specific differences from each estimated group mean to the estimated general mean. As required, the estimated differences add to zero and we find the group mean for $D3$ as $\hat{\mu}_3 = \hat{\mu} + \hat{\alpha}_3 = 11.43 + (-2.19) = 9.24$.

Using R

Base-R provides the `aov()` function for calculating an analysis of variance. The model is specified using R's formula framework, which implements a previously proposed symbolic description of models (Wilkinson and Rogers 1973). The `summary()` function prints the ANOVA table of a fitted model. Alternatively, the same model specification (but without `Error()`) can be used with the linear regression function `lm()` in which case `summary()` provides the (independent) regression parameter estimates and `dummy.coef()` gives a list of all parameters. The functions `summary.lm()` and `summary.aov()` provide the respective other view for these two equivalent models. Further details are provided in Sect. 4.5.3.

ANOVA tables can also be calculated from other types of regression models (we look at linear mixed models in subsequent chapters) using the function `anova()` with the fitted model. This function also allows specifying the approximation method for calculating degrees of freedom using its `ddf=` option. Power analysis for a one-way ANOVA is provided by `power.anova.test()`. Completely randomized experiments can be designed using `design.crd()` from package `agricolae`, which also provides randomization.

Exact confidence limits for f^2 and η^2 are translated from those calculated for the noncentrality parameter λ of the F-distribution. The `effectsize` package provides functions `cohens_f()` and `eta_squared()` for calculating these effect sizes and their confidence intervals.

Summary

The analysis of variance decomposes the overall variation in the data into parts attributable to different sources, such as treatment factors and residuals. The decomposition relies on a partition of the sum of squares, measuring the variation, and the degrees of freedom, measuring the amount of data expended on each source. The ANOVA table provides a summary of this decomposition: sums of squares and degrees of freedom for each source, and the resulting mean squares. The F-test tests the null hypothesis that all treatment group means are equal. It uses the ratio of the treatment means squares and the residual means squares, which provide two independent estimates of the residual variance under the null hypothesis. For two groups, this is equivalent to a t-test with equal group variances.

Hasse diagrams visualize the logical structure of an experiment: we distinguish the unit structure with the unit factors and their relations from the treatment structure with the treatment factors and their relations. We combine both into the experiment structure by linking each treatment factor to the unit factor on which it is randomized; this is the experimental unit for that treatment factor and provides the residual mean squares of the F-test. We derive the model specification directly from the Hasse diagram and can verify the correct specification by comparing the degrees of freedom from the diagram with those in the resulting ANOVA table.

References

Cochran, W. G. (1947). "Some Consequences When the Assumptions for the Analysis of Variance are not Satisfied". In: Biometrics 3.1, pp. 22–38.

Cohen, J. (1988). Statistical Power Analysis for the Behavioral Sciences. 2nd. Lawrence Erlbaum Associates, Hillsdale.

Cohen, J. (1992). "A Power Primer". In: Psychological Bulletin 112.1, pp. 155–159.

Cumming, G. and S. Finch (2001). "A primer on the understanding, use, and calculation of confidence intervals that are based on central and noncentral distributions". In: Educational and Psychological Measurement 61.4, pp. 532–574.

Fisher, R. A. (1934). "Discussion to "Statistics in Agricultural Research"". In: Journal of the Royal Statistical Society 1, pp. 26–61.

Julious, S. A. (2005). "Sample size of 12 per group rule of thumb for a pilot study". In: Pharmaceutical Statistics 4.4, pp. 287–291.

Kastenbaum, M. A., D. G. Hoel, and K. O. Bowman (1970). "Sample size requirements: One-way Analysis of Variance". In: Biometrika 57.2, pp. 421–430.

Kelley, K. (2007). "Confidence intervals for standardized effect sizes: theory, application, and implementation". In: Journal Of Statistical Software 20.8, e1–e24.

Kenward, M. G. and J. H. Roger (1997). "Small Sample Inference for Fixed Effects from Restricted Maximum Likelihood". In: Biometrics 53.3, pp. 983–997.

Kieser, M. and G. Wassmer (1996). "On the use of the upper confidence limit for the variance from a pilot sample for sample size determination". In: Biometrical Journal 38, pp. 941–949.

Kruskal, W. H. and W. A. Wallis (1952). "Use of Ranks in One-Criterion Variance Analysis". In: Journal of the American Statistical Association 47.260, pp. 583–621.

Lakens, D. (2013). "Calculating and reporting effect sizes to facilitate cumulative science: A practical primer for t-tests and ANOVAs". In: Frontiers in Psychology 4, pp. 1–12.

Lenth, R. V. (2016). "Least-Squares Means: The R Package lsmeans". In: Journal of Statistical Software 69.1.

Lindman, H. R. (1992). Analysis of Variance in Experimental Design. Springer Berlin/Heidelberg.

Nakagawa, S. and I. C. Cuthill (2007). "Effect size, confidence interval and statistical significance: a practical guide for biologists." In: Biological Reviews of the Cambridge Philosophical Society 82.4, pp. 591–605.

Nelder, J. A. (1994). "The statistics of linear models: back to basics". In: Statistics and Computing 4.4, pp. 221–234.

Nelder, J. A. (1998). "The great mixed-model muddle is alive and flourishing, alas!" In: Food Quality and Preference 9.3, pp. 157–159.

Satterthwaite, F. E. (1946). "An approximate distribution of estimates of variance components". In: Biometrika Bulletin 2.6, pp. 110–114.

Searle, S. R., F. M. Speed, and G. A. Milliken (1980). "Population Marginal Means in the Linear-Model - An Alternative to Least-Squares Means". In: American Statistician 34.4, pp. 216–221.

Tang, P. C. (1938). "The power function of the analysis of variance test with tables and illustrations of their use". In: Statistical Research Memoirs 2, pp. 126–149.

van Belle, G. (2008). Statistical Rules of Thumb. 2nd. John Wiley & Sons, Inc.

Venables, W. (1975). "Calculation of confidence intervals for noncentrality parameters". In: Journal of the Royal Statistical Society B 37.3, pp. 406–412.

Wheeler, R. E. (1974). "Portable Power". In: Technometrics 16.2, pp. 193–201.

Wilkinson, G. N. and C. E. Rogers (1973). "Symbolic description of factorial models for analysis of variance". In: Journal of the Royal Statistical Society C 22.3, pp. 392–399.

Chapter 5
Comparing Treatment Groups with Linear Contrasts

5.1 Introduction

The omnibus F-test appraises the evidence against the hypothesis of identical group means, but a rejection of this null hypothesis provides little information about *which* groups differ and *how*. A very general and elegant framework for evaluating treatment group differences are *linear contrasts*, which provide a principled way for constructing corresponding t-tests and confidence intervals. In this chapter, we develop this framework and apply it to our four drugs example; we also consider several more complex examples to demonstrate its power and versatility.

If a set of contrasts is *orthogonal*, then we can reconstitute the result of the F-test using the results from the contrasts, and a significant F-test implies that at least one contrast is significant. If the F-test is not significant, we might still find significant contrasts, because the F-test considers all deviations from equal group means simultaneously, while a contrast looks for a specific set of deviations for which it provides more power by ignoring other potential deviations.

Considering several contrasts leads to a *multiplicity problem* that requires adjusting confidence and significance levels. Several *multiple comparison procedures* allow us to calculate these adjustments for different sets of contrasts.

5.2 Linear Contrasts

For our example, we might be interested in comparing the two drugs $D1$ and $D2$, for example. One way of doing this is by a simple t-test between the corresponding observations. This yields a t-value of $t = 2.22$ and a p-value of $p = 0.044$ for a difference of $\hat{\mu}_1 - \hat{\mu}_2 = 1.52$ with a 95%-confidence interval [0.05, 2.98]. While this approach yields a valid estimate and test, it is inefficient because we completely neglect the information available in the observations of drugs $D3$ and $D4$. Specifi-

© Springer Nature Switzerland AG 2021
H.-M. Kaltenbach, *Statistical Design and Analysis of Biological Experiments*,
Statistics for Biology and Health, https://doi.org/10.1007/978-3-030-69641-2_5

cally, if we assume that the variances are the same in all treatment groups, we could use these additional observations to get a better estimate of the residual variance σ_e^2 and increase the degrees of freedom.

We consider three example comparisons using our four drugs. We additionally assume that $D1$ and $D2$ share the same active component and denote these drugs as 'Class A', while $D3$ and $D4$ share another component ('Class B'):

1. as before, compare the drugs in the first class: $D1$ versus $D2$;
2. compare the drugs in the second class: $D3$ versus $D4$;
3. compare the classes: average of $D1$ and $D2$ versus average of $D3$ and $D4$.

We can formulate these comparisons in terms of differences of treatment group means; each is an example of a linear contrast:

$$D1 \text{ versus } D2 : \mu_1 - \mu_2$$
$$D3 \text{ versus } D4 : \mu_3 - \mu_4$$
$$\text{Class A versus Class B} : \left(\frac{\mu_1 + \mu_2}{2}\right) - \left(\frac{\mu_3 + \mu_4}{2}\right) .$$

Note that a t-test for the third comparison requires manual calculation of the corresponding estimates and their standard errors first.

Linear contrasts use all data for estimation and 'automatically' lead to the correct t-test and confidence interval calculations. Their estimation is one of the main purposes for an experiment:

Contrasts of interest justify the design, not the other way around.

An important task in *designing* an experiment is to ensure that contrasts of interest are defined beforehand and can be estimated with adequate precision.

5.2.1 Defining Contrasts

Formally, a linear contrast $\Psi(\mathbf{w})$ for a treatment factor with k levels is a linear combination of the group means using a weight vector $\mathbf{w} = (w_1, \ldots, w_k)$:

$$\Psi(\mathbf{w}) = w_1 \cdot \mu_1 + \cdots + w_k \cdot \mu_k ,$$

where the entries in the weight vector sum to zero, such that $w_1 + \cdots + w_k = 0$.

We compare the group means of two sets X and Y of treatment factor levels by selecting the weights w_i as follows:

- the weight of each treatment level *not* considered is zero: $w_i = 0 \iff i \notin X$ and $i \notin Y$;
- the weights for set X are all positive: $w_i > 0 \iff i \in X$;

- the weights for set Y are all negative: $w_i < 0 \iff i \in Y$;
- the weights sum to zero: $w_1 + \cdots + w_k = 0$;
- the individual weights w_i determine how the group means of the sets X and Y are averaged; using equal weights with each set corresponds to a simple average of the set's group means.

The weight vectors for our example contrasts are $\mathbf{w}_1 = (+1, -1, 0, 0)$ for the first contrast, where $X = \{1\}$ and $Y = \{2\}$; $\mathbf{w}_2 = (0, 0, +1, -1)$ for the second, $X = \{3\}$ and $Y = \{4\}$; and $\mathbf{w}_3 = (+1/2, +1/2, -1/2, -1/2)$ for the third contrast, where $X = \{1, 2\}$ and $Y = \{3, 4\}$.

5.2.2 Estimating Contrasts

We estimate a contrast by replacing the group means by their estimates:

$$\hat{\Psi}(\mathbf{w}) = w_1 \cdot \hat{\mu}_1 + \cdots + w_k \cdot \hat{\mu}_k \, .$$

Unbalancedness only affects the precision but not the interpretation of contrast estimates, and we can make our exposition more general by allowing different numbers of samples per group, denoting by n_i the number of samples in group i. From the properties of the group mean estimates $\hat{\mu}_i$, we know that $\hat{\Psi}(\mathbf{w})$ is an unbiased estimator of the contrast $\Psi(\mathbf{w})$ and has variance

$$\mathrm{Var}\left(\hat{\Psi}(\mathbf{w})\right) = \mathrm{Var}\left(w_1 \cdot \hat{\mu}_1 + \cdots + w_k \cdot \hat{\mu}_k\right) = \sum_{i=1}^{k} w_i^2 \cdot \mathrm{Var}(\hat{\mu}_i) = \sigma_e^2 \cdot \sum_{i=1}^{k} \frac{w_i^2}{n_i} \, .$$

We can thus estimate its standard error by

$$\widehat{\mathrm{se}}\left(\hat{\Psi}(\mathbf{w})\right) = \hat{\sigma}_e \cdot \sqrt{\sum_{i=1}^{k} \frac{w_i^2}{n_i}} \, .$$

Note that the precision of a contrast estimate depends on the sizes for the involved groups (i.e., those with $w_i \neq 0$) in an unbalanced design, and standard errors are higher for contrasts involving groups with low numbers of replicates in this case.

The estimate of a contrast is based on the normally distributed estimates of the group means. We can use the residual variance estimate from the preceding ANOVA, and the resulting estimator for any contrast has a t-distribution with all $N - k$ degrees of freedom.

Table 5.1 Estimates and 95%-confidence intervals for three example contrasts

Comparison	Contrast	Estimate	LCL	UCL
$D1$-versus-$D2$	$\Psi(\mathbf{w}_1) = \mu_1 - \mu_2$	1.51	0.27	2.76
$D3$-versus-$D4$	$\Psi(\mathbf{w}_2) = \mu_3 - \mu_4$	-0.1	-1.34	1.15
Class A-versus-Class B	$\Psi(\mathbf{w}_3) = \frac{\mu_1+\mu_2}{2} - \frac{\mu_3+\mu_4}{2}$	4.28	3.4	5.16

This immediately yields a $(1 - \alpha)$-confidence interval for a contrast estimate $\hat{\Psi}(\mathbf{w})$:

$$\hat{\Psi}(\mathbf{w}) \pm t_{\alpha/2, N-k} \cdot \widehat{\mathrm{se}}\left(\hat{\Psi}(\mathbf{w})\right) = \hat{\Psi}(\mathbf{w}) \pm t_{\alpha/2, N-k} \cdot \hat{\sigma}_e \cdot \sqrt{\sum_{i=1}^{k} \frac{w_i^2}{n_i}} ,$$

where again $t_{\alpha/2, N-k}$ is the $\alpha/2$-quantile of the t-distribution with $N - k$ degrees of freedom (N the total number of samples). If the degrees of freedom are sufficiently large, we can alternatively calculate the confidence interval based on the normal quantiles by replacing the quantile $t_{\alpha/2, N-k}$ with $z_{\alpha/2}$.

For our third example contrast, we find an estimate of $\hat{\Psi}(\mathbf{w}_3) = 4.28$ for the difference between the average enzyme levels for a class A and class B drug. We already have an estimate $\hat{\sigma}_e = 1.22$ of the residual standard deviation: it is the root of the residual mean squares. The standard error of $\hat{\Psi}(\mathbf{w}_3)$ is given by

$$\widehat{\mathrm{se}}(\hat{\Psi}(\mathbf{w}_3)) = \hat{\sigma}_e \cdot \sqrt{\left(\frac{(0.5)^2}{8} + \frac{(0.5)^2}{8} + \frac{(-0.5)^2}{8} + \frac{(-0.5)^2}{8}\right)} \approx 0.35 \cdot \hat{\sigma}_e ,$$

which yields an estimated standard error of $\widehat{\mathrm{se}}(\hat{\Psi}(\mathbf{w}_3)) = 0.43$.

From this, we calculate a t-based 95%-confidence interval of $[3.4, 5.16]$, based on the t-quantile $t_{0.025,28} = -2.05$ for $N - k = 28$ degrees of freedom. The confidence interval only contains positive values and we can therefore conclude that drugs in class A have indeed a higher enzyme level than those in class B and the observed difference between the classes is not likely due to random fluctuations in the data.

Equivalent calculations for the remaining two example contrasts $\Psi(\mathbf{w}_1)$ and $\Psi(\mathbf{w}_2)$ yield the estimates and 95%-confidence intervals in Table 5.1. The results suggest that the two drugs $D1$ and $D2$ in class A lead to different enzyme levels, if only slightly so. Note that the estimate for this contrast is identical to our t-test result, but the confidence interval is substantially narrower; this is the result of pooling the data for estimating the residual variance and increasing the degrees of freedom. The two drugs in class B cannot be distinguished based on the data.

5.2.3 Testing Contrasts

A linear contrast estimate has a t-distribution for normally distributed response values. This allows us to derive a t-test for testing the null hypothesis

$$H_0 : \Psi(\mathbf{w}) = 0$$

using the test statistic

$$T = \frac{\hat{\Psi}(\mathbf{w})}{\widehat{\mathrm{se}}(\hat{\Psi}(\mathbf{w}))} = \frac{\sum_{i=1}^{k} w_i \hat{\mu}_i}{\hat{\sigma}_e \cdot \sqrt{\sum_{i=1}^{k} w_i^2 / n_i}},$$

which has a t-distribution with $N - k$ degrees of freedom if H_0 is true.

For our second example contrast, the t-statistic is

$$T = \frac{(0) \cdot \hat{\mu}_1 + (0) \cdot \hat{\mu}_2 + (+1) \cdot \hat{\mu}_3 + (-1) \cdot \hat{\mu}_4}{\hat{\sigma}_e \cdot \sqrt{\frac{(0)^2 + (0^2) + (+1)^2 + (-1)^2}{8}}} = \frac{\hat{\mu}_3 - \hat{\mu}_4}{\sqrt{2}\hat{\sigma}_e / \sqrt{8}}.$$

This is exactly the statistic for a two-sample t-test, but uses a pooled variance estimate over all treatment groups and $N - k = 28$ degrees of freedom. For our data, we calculate $t = -0.16$ and $p = 0.88$; the enzyme levels for drugs $D3$ and $D4$ cannot be distinguished. Equivalent calculations for all three example contrasts are summarized in Table 5.2. Note again that for our first contrast, the t-value is about 10% larger than with the t-test based on the two groups alone, and the corresponding p-value is about one-half.

Table 5.2 t-tests for three example contrasts

Contrast	Test	t value	se	$P(> |t|)$
$D1$-versus-$D2$	$H_0 : \mu_1 - \mu_2 = 0$	2.49	0.61	1.89e-02
$D3$-versus-$D4$	$H_0 : \mu_3 - \mu_4 = 0$	-0.16	0.61	8.75e-01
Class A-versus-Class B	$H_0 : \frac{\mu_1 + \mu_2}{2} - \frac{\mu_3 + \mu_4}{2} = 0$	9.96	0.43	1.04e-10

5.2.4 Using Contrasts in R

The emmeans package provides for the heavy lifting: we calculate the analysis of variance using aov(), estimate the group means using emmeans(), and define a list of contrasts which we estimate using contrast().

We are usually more interested in confidence intervals for contrast estimates than we are in t-values and test results. Conveniently, the confint() function takes a contrast() result directly and by default yields 95%-confidence intervals for our contrasts.

For our three example contrasts, the following code performs all required calculations in just five commands:

```
m = aov(y~drug, data=drugs)
em = emmeans(m, ~drug)
ourContrasts = list(
  "D1-vs-D2"=c(1,-1,0,0),
  "D3-vs-D4"=c(0,0,1,-1),
  "Class A-vs-Class B"=c(1/2,1/2,-1/2,-1/2)
  )
estimatedContrasts = contrast(em, method=ourContrasts)
ci = confint(estimatedContrasts)
```

5.2.5 Orthogonal Contrasts and ANOVA Decomposition

There is no limit on the number of contrasts that we might estimate for any set of data. On the other hand, contrasts are linear combinations of the k group means and we also need to estimate the grand mean.[1] That means that we can find exactly $k - 1$ contrasts that exhaust the information available in the group means and any additional contrast can be calculated from results of these $k - 1$ contrasts. This idea is made formal by saying that two contrasts $\Psi(\mathbf{w})$ and $\Psi(\mathbf{v})$ are *orthogonal* if

$$\sum_{i=1}^{k} \frac{w_i \cdot v_i}{n_i} = 0 .$$

This requirement reduces to the more interpretable 'usual' orthogonality condition $\sum_i w_i \cdot v_i = 0$ for a balanced design.

Our three example contrasts are all pair-wise orthogonal; for example, we have $\sum_i w_{1,i} \cdot w_{2,i} = (+1 \cdot 0) + (-1 \cdot 0) + (0 \cdot +1) + (0 \cdot -1) = 0$ for the first and sec-

---

[1] We can do this using a 'contrast' $\mathbf{w} = (1/k, 1/k, \ldots, 1/k)$, even though its weights do not sum to zero.

ond contrast. With $k = 4$, only three contrasts can be mutually orthogonal and our three contrasts thus fully exhaust the available information.

If two contrasts $\Psi(\mathbf{w})$ and $\Psi(\mathbf{v})$ are orthogonal, then the associated null hypotheses

$$H_0 : \Psi(\mathbf{w}) = 0 \quad \text{and} \quad H_0 : \Psi(\mathbf{v}) = 0$$

are logically independent in the sense that we can learn nothing about one being true or false by knowing the other being true or false.[2]

A set of $k - 1$ orthogonal contrasts decomposes the treatment sum of squares into $k - 1$ *contrast sums of squares*. The sum of squares of a contrast $\Psi(\mathbf{w})$ is

$$SS_{\mathbf{w}} = \frac{\left(\sum_{i=1}^{k} w_i \hat{\mu}_i\right)^2}{\sum_{i=1}^{k} w_i^2 / n_i},$$

and each contrast has one degree of freedom: $df_{\mathbf{w}} = 1$. We can use the F-statistic

$$F = \frac{SS_{\mathbf{w}}/1}{\hat{\sigma}_e^2} = \frac{SS_{\mathbf{w}}}{MS_{\text{res}}}$$

with 1 numerator and $N - k$ denominator degrees of freedom for testing $H_0 : \Psi(\mathbf{w}) = 0$; this is equivalent to the t-test.

For our example contrasts, we find $SS_{\mathbf{w}_1} = 9.18$, $SS_{\mathbf{w}_2} = 0.04$, and $SS_{\mathbf{w}_3} = 146.68$ with associated F-values $F_{\mathbf{w}_1} = 6.21$, $F_{\mathbf{w}_2} = 0.03$, and $F_{\mathbf{w}_3} = 99.27$, each with 1 numerator and 28 denominator degrees of freedom. Note that $\sqrt{F_{\mathbf{w}_1}} = 2.49$ corresponds precisely to the absolute value of the previous t-statistic in Table 5.2, for example.

We can reconstitute the omnibus F-test for the treatment factor from the contrast F-tests. In particular, we know that if the omnibus F-test is significant, so is at least one of the contrasts in an orthogonal set. For our example, we find the treatment sum of squares as $SS_{\mathbf{w}_1} + SS_{\mathbf{w}_2} + SS_{\mathbf{w}_3} = 155.89$, and the treatment F-value as $(F_{\mathbf{w}_1} + F_{\mathbf{w}_2} + F_{\mathbf{w}_3})/3 = 35.17$. These results correspond exactly to the values from our previous ANOVA (Table 4.2).

Orthogonal contrasts provide a systematic way to ensure that the data of an experiment are fully exhausted in the analysis. In practice, scientific questions are sometimes more directly addressed by sets of non-orthogonal contrasts which are then preferred for their easier interpretation, even though their hypothesis tests might be logically dependent and contain redundancies.

---

[2] A small subtlety arises from estimating the contrasts: since all t-tests are based on the same estimate of the residual variance, the tests are still statistically dependent. The effect is usually so small that we ignore this subtlety in practice.

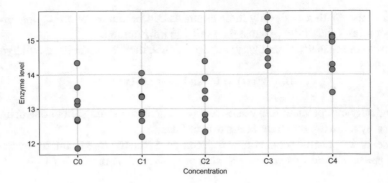

Fig. 5.1   Enzyme levels for five increasing concentrations of drug D1

5.2.6   Contrasts for Ordered Factors

The order of levels of **Drug** is completely arbitrary: we could just as well put the drugs of Class B as the first two levels and those of Class A as levels three and four. For some treatment factors, levels are *ordered* and level i is 'smaller' than level $i + 1$ in some well-defined sense; such factors are called *ordinal*. For example, our treatment might consist of our current drug $D1$ administered at equally spaced concentrations $C_0 < C_1 < C_2 < C_3 < C_4$. Data for 8 mice per concentration are shown in Fig. 5.1 and indicate that the drug's effect is negligible for the first two or three concentrations, then increases substantially and seems to decrease again for higher concentrations.

Trend Analysis Using Orthogonal Polynomials

We can analyze such data using ordinary contrasts, but with ordered treatment factor levels, it makes sense to look for *trends*. We do this using *orthogonal polynomials*, which are linear, quadratic, cubic, and quartic polynomials that decompose a potential trend into different components. Orthogonal polynomials are formulated as a special set of orthogonal contrasts. We use the emmeans() and contrast() combination again, and ask for a poly contrast to generate the appropriate set of orthogonal contrasts:

```
m.conc = aov(y~concentration, data=drug_concentrations)
em.conc = emmeans(m.conc, ~concentration)
ct.conc = contrast(em.conc, method="poly")
```

The contrast weight vectors for our example are shown in Table 5.3.

Each polynomial contrast measures the similarity of the shape of the data to the pattern described by the weight vector. The linear polynomial measures an upward or downward trend, while the quadratic polynomial measures curvature in the response

Table 5.3 Weight vectors for contrasts of orthogonal polynomials

	Linear	Quadratic	Cubic	Quartic
C0	−2	2	−1	1
C1	−1	−1	2	−4
C2	0	−2	0	6
C3	1	−1	−2	−4
C4	2	2	1	1

Table 5.4 Estimates for polynomial contrasts in drug concentration example

Contrast	Estimate	se	df	t value	P(>|t|)
Linear	4.88	0.70	35	6.95	4.38e-08
Quadratic	0.50	0.83	35	0.61	5.48e-01
Cubic	−2.14	0.70	35	−3.05	4.40e-03
Quartic	−5.20	1.86	35	−2.80	8.32e-03

to the concentrations, such that the trend is not simply a proportional increase or decrease in enzyme level. Cubic and quartic polynomials measure more complex curvature, but become harder to interpret directly.

For our data, we get the result in Table 5.4. We find a highly significant positive linear trend, which means that on average, the enzyme level increases with increasing concentration. The negligible quadratic together with significant cubic and quartic trend components means that there is curvature in the data, but it is changing with the concentration. This reflects the fact that the data show a large increase for the fourth concentration, which then levels off or decreases at the fifth concentration, leading to a sigmoidal pattern.

Time Trends

Another typical example of an ordered factor is time. We have to be careful about the experimental design, however. For example, measuring the enzyme levels at four timepoints using different mice per timepoint can be analyzed using orthogonal polynomials and an analysis of variance approach. This is because each mouse can be randomly allocated to one timepoint, yielding a completely randomized design. If, however, the design is such that the *same* mice are measured at four different timepoints, then the assumption of random assignment is violated and analysis of variance with contrasts is no longer a reasonable analysis option.[3] This type of *longitudinal* design, where the same unit is followed over time, requires more specialized

---

[3] It is plausible that measurements on the same mouse are more similar between timepoints close together than between timepoints further apart, a fact that ANOVA cannot properly capture.

methods to capture the inherent correlation between timepoints. We briefly discuss simple longitudinal designs in Sect. 8.4.

Minimal Effective Dose

Another example of contrasts useful for ordered factors are the orthogonal *Helmert contrasts*. They compare the second level to the first, the third level to the average of the first and second level, the fourth level to the average of the preceding three, and so forth.

Helmert contrasts can be used for finding a *minimal effective dose* in a dose-response study. Since doses are typically in increasing order, we first test the second-lowest against the lowest dose. If the corresponding average responses cannot be distinguished, we assume that no effect of relevant size is present (provided the experiment is not underpowered). We then pool the data of these two doses and use their common average to compare against the third-smallest dose, thereby increasing the precision compared to contrasting each level only to its preceding level.

Helmert contrasts are not directly available in emmeans, but the package manual tells us how to define them ourselves:

```
helmert.emmc = function(ordered.levels, ...) {
    # Use built-in R contrast to find contrast matrix
    contrast.matrix = as.data.frame(contr.helmert(ordered.levels))
    # Provide useful name for each contrast
    names(contrast.matrix) = paste(ordered.levels[-1],"vs lower")
    # Provide name of contrast set
    attr(contrast.matrix, "desc") = "Helmert contrasts"
    return(contrast.matrix)
}
```

In our example data in Fig. 5.1, the concentration C_3 shows a clear effect. The situation is much less clear-cut for lower concentrations $C_0, \ldots, C_2$; there might be a hint of linear increase, but this might be due to random fluctuations. Using the Helmert contrasts

```
ct.helmert = contrast(em.conc, "helmert")
```

yields the contrasts in Table 5.5 and the results in Table 5.6. Concentrations C_0, C_1, C_2 show no discernible differences, while enzyme levels increase significantly for concentration C_3, indicating that the minimal effective dose is between concentrations C_2 and C_3.

Table 5.5 Weight vectors for orthogonal Helmert contrasts

Concentration	C1 versus lower	C2 versus lower	C3 versus lower	C4 versus lower
C0	-1	-1	-1	-1
C1	1	-1	-1	-1
C2	0	2	-1	-1
C3	0	0	3	-1
C4	0	0	0	4

Table 5.6 Estimates for Helmert contrasts indicate minimum effective dose of at most C3

Contrast	Estimate	se	df	t value	P(>|t|)
C1 versus lower	0.11	0.31	35	0.35	7.28e-01
C2 versus lower	0.38	0.54	35	0.71	4.84e-01
C3 versus lower	5.47	0.77	35	7.11	2.77e-08
C4 versus lower	3.80	0.99	35	3.83	5.10e-04

5.2.7 Standardized Effect Size

Similar to our previous discussions for simple differences and omnibus F-tests, we might sometimes profit from a standardized effect size measure for a linear contrast $\Psi(\mathbf{w})$, which provides the size of the contrast in units of standard deviation.

A first idea is to directly generalize Cohen's d as $\Psi(\mathbf{w})/\sigma_e$ and measure the contrast estimate in units of the standard deviation. A problem with this approach is that the measure still depends on the weight vector: if $\mathbf{w} = (w_1, \ldots, w_k)$ is the weight vector of a contrast, then we can define an equivalent contrast using, for example, $\mathbf{w}' = 2 \cdot \mathbf{w} = (2 \cdot w_1, \ldots, 2 \cdot w_k)$. Then, $\Psi(\mathbf{w}') = 2 \cdot \Psi(\mathbf{w})$, and the above standardized measure also scales accordingly. In addition, we would like our standardized effect measure to equal Cohen's d if the contrast describes a simple difference between two groups.

Both problems are resolved by Abelson's standardized effect size measure $d_{\mathbf{w}}$ (Abelson and Prentice 1997):

$$d_{\mathbf{w}} = \sqrt{\frac{2}{\sum_{i=1}^{k} w_i^2}} \cdot \frac{\Psi(\mathbf{w})}{\sigma_e} \text{ estimated by } \hat{d}_{\mathbf{w}} = \sqrt{\frac{2}{\sum_{i=1}^{k} w_i^2}} \cdot \frac{\hat{\Psi}(\mathbf{w})}{\hat{\sigma}_e}.$$

For our scaled contrast $\mathbf{w}'$, we find $d_{\mathbf{w}'} = d_{\mathbf{w}}$ and the standardized effect sizes for $\mathbf{w}'$ and $\mathbf{w}$ coincide. For a simple difference $\mu_i - \mu_j$, we have $w_i = +1$ and $w_j = -1$, all other weights being zero. Thus $\sum_{i=1}^{k} w_i^2 = 2$ and $d_{\mathbf{w}}$ is reduced to Cohen's d.

5.2.8 Power Analysis and Sample Size

The power calculations for linear contrasts can be done based on the contrast estimate and its standard error, which requires calculating the power from the noncentral t-distribution. We follow the equivalent approach based on the contrast's sum of squares and the residual variance, which requires calculating the power from the noncentral F-distribution.

Exact Method

The noncentrality parameter λ of the F-distribution for a contrast is given as

$$\lambda = \frac{\Psi(\mathbf{w})^2}{\text{Var}(\hat{\Psi}(\mathbf{w}))} = \frac{\left(\sum_{i=1}^{k} w_i \cdot \mu_i\right)^2}{\frac{\sigma^2}{n} \sum_{i=1}^{k} w_i^2} = n \cdot 2 \cdot \left(\frac{d_\mathbf{w}}{2}\right)^2 = n \cdot 2 \cdot f_\mathbf{w}^2.$$

Note that the last term has exactly the same $\lambda = n \cdot k \cdot f^2$ form that we encountered previously, since a contrast uses $k = 2$ (sets of) groups and we know that $f^2 = d^2/4$ for direct group comparisons.

From the noncentrality parameter, we can calculate the power for testing the null hypothesis $H_0 : \Psi(\mathbf{w}) = 0$ for any given significance level α, residual variance σ_e^2, sample size n, and assumed real value of the contrast Ψ_0, respectively, the assumed standardized effect size $d_{\mathbf{w},0}$. We calculate the power using our getPowerF() function and increase n until we reach the desired power.

For our first example contrast $\mu_1 - \mu_2$, we find $\sum_i w_i^2 = 2$. For a minimal difference of $\Psi_0 = 2$ and using a residual variance estimate $\hat{\sigma}_e^2 = 1.5$, we calculate a noncentrality parameter of $\lambda = 1.33\,n$. The numerator degree of freedom is df1=1 and the denominator degrees of freedom are df2=n*4-4. For a significance level of $\alpha = 5\%$ and a desired power of $1 - \beta = 80\%$, we find a required sample size of $n = 7$. We arrive at a very conservative estimate by replacing the residual variance by the upper confidence limit UCL $= 2.74$ of its estimate. This increases the required sample size to $n = 12$. The sample size increases substantially to $n = 95$ if we want to detect a much smaller contrast value of $\Psi_0 = 0.5$.

Similarly, our third example contrast $(\mu_1 + \mu_2)/2 - (\mu_3 + \mu_4)/2$ has $\sum_i w_i^2 = 1$. A minimal value of $\Psi_0 = 2$ can be detected with 80% power at a 5% significance level for a sample size of $n = 4$ per group, with an exact power of 85% (for $n = 3$ the power is 70%). Even though the desired minimal value is identical to the first example contrast, we need less samples since we are comparing the average of two groups to the average of two other groups, making the estimate of the difference more precise.

The overall experiment size is then the maximum sample size required for any contrast of interest.

Portable Power

For making the power analysis for linear contrasts portable, we apply the same ideas as for the omnibus F-test. The numerator degrees of freedom for a contrast F-test is $df_{num} = 1$, and we find a sample size formula of

$$n = \frac{2 \cdot \phi^2 \cdot \sigma_e^2 \cdot \sum_{i=1}^{k} w_i^2}{\Psi_0^2} = \frac{\phi^2}{(d_w/2)^2} = \frac{\phi^2}{f_w^2}.$$

For a simple difference contrast, $\sum_i w_i^2 = 2$, we have $\Psi_0 = \delta_0$. With the approximation $\phi^2 \approx 4$ for a power of 80% at a 5% significance level and $k = 2$ factor levels, we derive our old formula again: $n = 16/(\Psi_0/\sigma_e)^2$.

For our first example contrast with $\Psi_0 = 2$, we find an approximate sample size of $n \approx 6$ based on $\phi^2 = 3$, in reasonable agreement with the exact sample size of $n = 7$. If we instead ask for a minimal difference of $\Psi_0 = 0.5$, this number increases to $n \approx 98$ mice per drug treatment group (exact: $n = 95$).

The sample size is lower for a contrast between two averages of group means. For our third example contrast with weights $\mathbf{w} = (1/2, 1/2, -1/2, -1/2)$, we find $n = 8/(\Psi_0/\sigma_e)^2$. With $\Psi_0 = 2$, this gives an approximate sample size of $n \approx 5$ (exact: 4).

5.3 Multiple Comparisons and Post-Hoc Analyses

5.3.1 Introduction

Planned and Post-hoc Contrasts

The use of orthogonal contrasts—pre-defined before the experiment is conducted— provides no further difficulty in the analysis. They are, however, often inconvenient for an informative analysis: we may want to use more than $k - 1$ *planned contrasts* as part of our pre-defined analysis strategy. In addition, exploratory experiments lead to *post-hoc contrasts* that are based on observed outcomes. Post-hoc contrasts also play a role in carefully planned and executed confirmatory experiments, when peculiar patterns emerge that were not anticipated, but require further investigation.

We must carefully distinguish between planned and post-hoc contrasts: the first case just points to the limitations of independent pre-defined contrasts, but the contrasts and hypotheses are still pre-defined at the planning stage of the experiment and *before* the data is in. In the second case, we 'cherry-pick' interesting results from the data *after* the experiment is done and we inspected the results. This course of action increases our risk of finding false positives. A common example of a post-hoc contrast occurs when looking at all pair-wise comparisons of treatments and defining the difference between the largest and smallest group mean as a contrast of interest.

Since there is *always* one pair with greatest difference, it is incorrect to use a standard
t-test for this contrast. Rather, the resulting p-value needs to be adjusted for the fact
that we cherry-picked our contrast, making it larger on average than the contrast of
any randomly chosen pair. Properly adjusted tests for post-hoc contrasts thus have
lower power than those of pre-defined planned contrasts.

The Problem of Multiple Comparisons

Whether we are testing several pre-planned contrasts or post-hoc contrasts of some
kind, we also have to adjust our analysis for *multiple comparisons*.

Imagine a scenario where we are testing q hypotheses, from q contrasts, say.
We want to control the false positive probability using a significance level of α for
each individual test. The probability that any individual test falsely rejects the null
hypothesis of a zero difference is then α. However, even if all q null hypotheses
are true, the probability that *at least one* of the tests will incorrectly reject its null
hypothesis is *not* α, but rather $1 - (1 - \alpha)^q$.

For $q = 200$ contrasts and a significance level of $\alpha = 5\%$, for example, the probability of at least one incorrect rejection is 99.9965%, a near certainty.[4] Indeed,
the expected number of false positives, given that all 200 null hypotheses are true,
is $200 \cdot \alpha = 10$. Even if we only test $q = 5$ hypotheses, the probability of at least
one incorrect rejection is 22.6% and thus substantially larger than the desired false
positive probability.

The probability of at least one false positive in a family of tests (the *family-wise*
or *experiment-wise* error probability) increases with the number of tests, and in any
case exceeds the individual test's significance level α. This is known as the *multiple
testing problem*. In essence, we have to decide which error we want to control: is it
the individual error per hypothesis or is it the overall error of at least one incorrect
declaration of significance in the whole family of hypotheses?

A *multiple comparison procedure (MCP)* is an algorithm that computes adjusted
significance levels for each individual test, such that the overall error probability is
bound by a pre-defined probability threshold. Some of the procedures are universally
applicable, while others are predicated on specific classes of contrasts (such as comparing all pairs), but offer higher power than more general procedures. Specifically,
(i) the universal *Bonferroni* and *Bonferroni–Holm corrections* provide control for all
scenarios of pre-defined hypotheses; (ii) *Tukey's honest significant difference (HSD)*
provides control for testing all pairs of groups; (iii) *Dunnett's method* covers the
case of comparing each group to a common control group; and (iv) *Scheffé's method*
gives confidence intervals and tests for any post-hoc comparisons suggested by the
data.

---

[4]If 200 hypotheses seem excessive, consider a simple microarray experiment: here, the difference
in expression level is simultaneously tested for thousands of genes.

These methods apply generally for multiple hypotheses. Here, we focus on testing q contrasts $\Psi(\mathbf{w}_l)$ with q null hypotheses

$$H_{0,l} : \Psi(\mathbf{w}_l) = 0,$$

where $\mathbf{w}_l = (w_{1l}, \ldots, w_{kl})$ is the weight vector describing the lth contrast.

5.3.2 General Purpose: Bonferroni–Holm

The Bonferroni and Holm corrections are popular and simple methods for controlling the family-wise error probability. Both work for arbitrary sets of planned contrasts, but are conservative and lead to low significance levels for the individual tests, often much lower than necessary.

The *simple Bonferroni method* is a *single-step* procedure to control the family-wise error probability by adjusting the individual significance level from α to α/q. It does not consider the observed data and rejects the null hypothesis $H_0 : \Psi(\mathbf{w}_l) = 0$ if the contrast exceeds the critical value based on the adjusted t-quantile:

$$\left| \hat{\Psi}(\mathbf{w}_l) \right| > t_{1-\alpha/2q, N-k} \cdot \hat{\sigma}_e \cdot \sqrt{\sum_{i=1}^{k} \frac{w_{il}^2}{n_i}} .$$

It is easily applied to existing test results: just multiply the original p-values by the number of tests q and declare a test significant if this adjusted p-value stays below the original significance level α.

For our three example contrasts, we previously found unadjusted p-values of 0.019 for the first, 0.875 for the second, and 10^{-10} for the third contrast. The Bonferroni adjustment consists of multiplying each by $q = 3$, resulting in adjusted p-values of 0.057 for the first, 2.626 for the second (which we cap at 1.0), and 3×10^{-10} for the third contrast, moving the first contrast from significant to not significant at the $\alpha = 5\%$ level. The resulting contrast estimates and t-test are shown in Table 5.7.

The *Bonferroni–Holm* method is based on the same assumptions, but uses a *multi-step* procedure to find an optimal significance level based on the observed data. This increases its power compared to the simple procedure. Let us call the unadjusted p-values of the q hypotheses $p_1, \ldots, p_q$. The method first sorts the observed p-value

Table 5.7 Estimated contrasts and hypothesis tests adjusted using Bonferroni correction

Contrast	Estimate	se	df	t value	P(>|t|)
D1-versus-D2	1.51	0.61	28	2.49	5.66e-02
D3-versus-D4	−0.10	0.61	28	−0.16	1.00e+00
Class A-versus-Class B	4.28	0.43	28	9.96	3.13e-10

such that $p_{(1)} < p_{(2)} < \cdots < p_{(q)}$ and $p_{(i)}$ is the ith smallest observed p-value. It then compares $p_{(1)}$ to α/q, $p_{(2)}$ to $\alpha/(q-1)$, $p_{(3)}$ to $\alpha/(q-2)$ and so on until a p-value exceeds its corresponding threshold. This yields the smallest index j such that

$$p_{(j)} > \frac{\alpha}{q+1-j} \, ,$$

and any hypothesis $H_{0,i}$ for which $p_i < p_{(j)}$ is declared significant.

5.3.3 Comparisons of All Pairs: Tukey

We gain more power if the set of contrasts has more structure, and Tukey's method is designed for the common case of all *pair-wise differences*. It considers the distribution of the *studentized range*, the difference between the maximal and minimal group means by calculating *honest significant differences (HSD)* (Tukey 1949a). It requires a balanced design and rejects $H_{0,l} : \Psi(\mathbf{w}_l) = 0$ if

$$\left| \hat{\Psi}(\mathbf{w}_l) \right| > q_{1-\alpha,k-1,N} \cdot \hat{\sigma}_e \cdot \sqrt{\frac{1}{2} \sum_{i=1}^{k} \frac{w_{il}^2}{n}} \quad \text{that is} \quad |\hat{\mu}_i - \hat{\mu}_j| > q_{1-\alpha,k-1,N} \cdot \frac{\hat{\sigma}_e}{\sqrt{n}} \, ,$$

where $q_{\alpha,k-1,N}$ is the α-quantile of the studentized range based on k groups and $N = n \cdot k$ samples.

The result is shown in Table 5.8 for our example. Since the difference between the two drug classes is very large, all but the two comparisons within each class yield highly significant estimates, but neither difference of drugs in the same class is significant after Tukey's adjustment.

5.3.4 Comparisons Against a Reference: Dunnett

Another very common type of biological experiment uses a control group and inference focuses on comparing each treatment with this control group, leading to $k-1$ contrasts. These contrasts are not orthogonal, and the required adjustment is provided by Dunnett's method (Dunnett 1955). It rejects the null hypothesis $H_{0,i} : \mu_i - \mu_1 = 0$ that treatment group i shows no difference to the control group 1 if

$$|\hat{\mu}_i - \hat{\mu}_1| > d_{1-\alpha,k-1,N-k} \cdot \hat{\sigma}_e \cdot \sqrt{\frac{1}{n_i} + \frac{1}{n_1}} \, ,$$

where $d_{1-\alpha,k-1,N-k}$ is the quantile of the appropriate distribution for this test.

Table 5.8 Estimated contrasts of all pair-wise comparisons adjusted by Tukey's method (top) and versus the reference adjusted using Dunnett's method (bottom). Note that contrast $D1 - D2$, for example, yields identical estimates but different p-values

Contrast	Estimate	se	df	t value	P(>|t|)
Pairwise-Tukey					
D1 – D2	1.51	0.61	28	2.49	8.30e-02
D1 – D3	5.09	0.61	28	8.37	2.44e-08
D1 – D4	4.99	0.61	28	8.21	3.59e-08
D2 – D3	3.57	0.61	28	5.88	1.45e-05
D2 – D4	3.48	0.61	28	5.72	2.22e-05
D3 – D4	−0.10	0.61	28	−0.16	9.99e-01
Reference-Dunnett					
D2 – D1	−1.51	0.61	28	−2.49	5.05e-02
D3 – D1	−5.09	0.61	28	−8.37	1.24e-08
D4 – D1	−4.99	0.61	28	−8.21	1.82e-08

For our example, let us assume that drug $D1$ is the best current treatment option, and we are interested in comparing the alternatives $D2$ to $D4$ to this reference. The required contrasts are the differences from each drug to the reference $D1$, resulting in Table 5.8.

Unsurprisingly, enzyme levels for drug $D2$ are barely distinguishable from those of the reference drug $D1$, and $D3$ and $D4$ show very different responses than the reference.

5.3.5 General Purpose and Post-Hoc Contrasts: Scheffé

The method by Scheffé is suitable for testing any group of contrasts, even if they were suggested by the data (Scheffé 1959); in contrast, most other methods are restricted to pre-defined contrasts. Naturally, this freedom of cherry-picking comes at a cost: the Scheffé method is extremely conservative (so effects have to be huge to be deemed significant), and is therefore only used if no other method is applicable.

The Scheffé method rejects the null hypothesis $H_{0,l}$ if

$$\left| \hat{\Psi}(\mathbf{w}_l) \right| > \sqrt{(k-1) \cdot F_{\alpha,k-1,N-k}} \cdot \hat{\sigma}_e \cdot \sqrt{\sum_{i=1}^{k} \frac{w_{il}^2}{n_i}} .$$

This is very similar to the Bonferroni correction, except that the number of contrasts q is irrelevant, and the quantile is a scaled quantile of an F-distribution rather than a quantile from a t-distribution.

Table 5.9 Estimated contrasts of our three example contrasts, assuming they were suggested by the data; adjusted using Scheffe's method

Contrast	Estimate	se	df	t value	P(>|t|)
D1-versus-D2	1.51	0.61	28	2.49	1.27e-01
D3-versus-D4	−0.10	0.61	28	−0.16	9.99e-01
Class A-versus-Class B	4.28	0.43	28	9.96	2.40e-09

For illustration, imagine that our three example contrasts were not planned before the experiment, but rather suggested by the data after the experiment was completed. The adjusted results are shown in Table 5.9.

We notice that the *p*-values are much more conservative than with any other method, which reflects the added uncertainty due to the post-hoc nature of the contrasts.

5.3.6 Remarks

There is sometimes disagreement on when multiple comparison corrections are necessary and how strict or conservative they should be. In general, however, results not adjusted for multiple comparisons are unlikely to be accepted without further elaboration. Note that if multiple comparison procedures are required for multiple testing, we also need to adjust quantiles in confidence intervals accordingly.

On the one extreme, it is relatively uncontroversial that sets of post-hoc contrasts always require an appropriate adjustment of the significance levels to achieve reasonable credibility of the analysis. On the other extreme, planned orthogonal contrasts simply partition the treatment factor sum of squares and their hypotheses are independent, so they are usually not adjusted.

Many reasonable sets of contrasts are more nuanced than "every pair" or "everyone against the reference" but still have more structure than an arbitrary set of contrasts. Specialized methods might not exist, while general methods such as Bonferroni–Holm are likely more conservative than strictly required. The analyst's discretion is then required and the most important aspect is to honestly report what was done, such that others can fully appraise the evidence in the data.

5.4 A Real-Life Example—Drug Metabolization

We further illustrate the one-way ANOVA and use of linear contrasts using a real-life example (Lohasz et al. 2020).[5] The two anticancer drugs cyclophosphamide (CP) and ifosfamide (IFF) become active in the human body only after metabolization in the liver by the enzyme CYP3A4, among others. The function of this enzyme is inhibited by the drug ritanovir (RTV), which more strongly affects metabolization of IFF than CP. The experimental setup consisted of 18 independent channels distributed over several microfluidic devices; each channel contained a co-culture of multi-cellular liver spheroids for metabolization and tumor spheroids for measuring drug action.

The experiment used the diameter of the tumor (in μm) after 12 days as the response variable. There are six treatment groups: a control condition without drugs, a second condition with RTV alone, and the four conditions CP-only, IFF-only, and the combined CP:RTV and IFF:RTV. The resulting data are shown in Fig. 5.2A for each channel.

A preliminary analysis revealed that device-to-device variation and variation from channel to channel were negligible compared to the within-channel variance, and these two factors were consequently ignored in the analysis. Thus, data are pooled over the channels for each treatment group, and the experiment is analyzed as a (slightly unbalanced) one-way ANOVA. We discuss an alternative two-way ANOVA in Sect. 6.3.8.

Inhomogeneous Variances

The omnibus F-test and linear contrast analysis require equal within-group variances between treatment groups. This hypothesis was tested using the Levene test and a p-value below 0.5% indicated that variances might differ substantially. If true, this would complicate the analysis. Looking at the raw data in Fig. 5.2A, however, reveals a potential error in channel 16, which was labeled as IFF:RTV, but shows tumor diameters in excellent agreement with the neighboring CP:RTV treatment. Including channel 16 in the IFF:RTV group then inflates the variance estimate. It was therefore decided to remove channel 16 from further analysis, the hypothesis of equal variances is then no longer rejected ($p > 0.9$), and visual inspection of the data confirms that dispersions are very similar between channels and groups.

---

[5]The authors of this study kindly granted permission to use their data. Purely for illustration, we provide some alternative analyses to those in the publication.

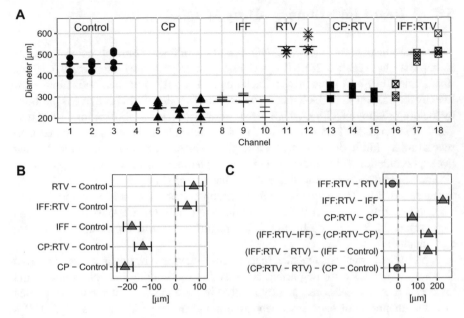

Fig. 5.2 **A** Observed diameters by channel. Point shape indicates treatment. Channel 16 appears to be mislabelled. **B** Estimated treatment-versus-control contrasts and Dunnett-adjusted 95%-confidence intervals based on data excluding channel 16, significant contrasts are indicated as triangles. **C** As (B) for specific (unadjusted) contrasts of interest

Analysis of Variance

As is expected from looking at the data in Fig. 5.2A, the one-way analysis of variance of tumor diameter versus treatment results in a highly significant treatment effect.

	Df	Sum Sq	Mean Sq	F value	Pr(>F)
Condition	5	815204.9	163040.98	154.53	6.34e-33
Residuals	60	63303.4	1055.06		

The effect sizes are an explained variation of $\eta^2 = 93\%$, a standardized effect size of $f^2 = 12.88$, and a raw effect size of $b^2 = 13589.1 \mu m^2$ (an average deviation between group means and general mean of $b = 116.57 \mu m$).

Linear Contrasts

Since the F-test does not elucidate which groups differ and by how much, we proceed with a more targeted analysis using linear contrasts to estimate and test meaningful and interpretable comparisons. With a first set of standard contrasts, we compare each treatment group to the control condition. The resulting contrast estimates are shown in Fig. 5.2B together with their Dunnett-corrected 95%-confidence intervals.

Somewhat surprisingly, the RTV-only condition shows tumor diameters significantly *larger* than those under the control condition, indicating that RTV alone influences the tumor growth. Both conditions involving CP show reduced tumor diameters, indicating that CP inhibits tumor growth, as does IFF alone. Lastly, RTV seems to substantially decrease the efficacy of IFF, leading again to tumor diameters larger than under the control condition, but (at least visually) comparable to the RTV condition.

The large and significant difference between control and RTV-only poses a problem for the interpretation: we are interested in comparing CP:RTV against CP-only and similarly for IFF. But CP:RTV could be a combined effect of tumor diameter reduction by CP (compared to control) and increase by RTV (compared to control). We have two options for defining a meaningful contrast: (i) estimate the difference in tumor diameter between CP:RTV and CP. This is a comparison between the combined and single drug actions. Or (ii) estimate the difference between the change in tumor diameter from CP to control and the change from CP:RTV to RTV (rather than to control). This is a comparison between the baseline tumor diameters under control and RTV to those under addition of CP and is the net-effect of CP (provided that RTV increases tumor diameters equally with and without CP).

The two sets of comparisons lead to different contrasts, but both are meaningful for these data. The authors of the study decided to go for the first type of comparison and compared tumor diameters for each drug with and without inhibitor. The two contrasts are IFF:RTV − IFF and CP:RTV − CP, shown in rows 2 and 3 of Fig. 5.2C. Both contrasts show a large and significant increase in tumor diameter in the presence of the inhibitor RTV, where the larger loss of efficacy for IFF yields a more pronounced difference.

For a complete interpretation, we are also interested in comparing the two differences between the CP and the IFF conditions: is the reduction in tumor diameter under CP smaller or larger than under IFF? This question addresses a *difference of differences*, a very common type of comparison in biology, when different conditions are contrasted and a 'baseline' or control is available for each condition. We express this question as (IFF:RTV − IFF) − (CP:RTV − CP), and we can sort the terms to derive the contrast form (IFF:RTV + CP) − (CP:RTV + IFF). Thus, we use weights of +1 for the treatment groups IFF:RTV and CP, and weights of −1 for the groups CP:RTV and IFF to define the contrast. Note that this differs from our previous example contrast comparing drug classes, where we compared averages of several groups by using weights ±1/2. The estimated contrast and confidence interval is shown in the fourth row in Fig. 5.2C: the two increases in tumor diameter under co-administered RTV are significantly different, with about 150 μm more under IFF than under CP.

For comparison, the remaining two rows in Fig. 5.2C show the tumor diameter increase for each drug with and without inhibitor, where the no-inhibitor condition is compared to the control condition, but the inhibitor condition is compared to the RTV-only condition. Now, IFF shows less increase in tumor diameter than in the previous comparison, but the result is still large and significant. In contrast, we do not find a difference between the CP and CP:RTV conditions indicating that loss of efficacy for CP is only marginal under RTV. This is because the previously observed difference can be explained by the difference in 'baseline' between control and RTV-only. The contrasts are constructed as before: for CP, the comparison is (CP:RTV − RTV) − (CP − Control) which is equivalent to (CP:RTV + Control) − (RTV + CP).

Conclusion

What started as a seemingly straightforward one-way ANOVA turned into a much more intricate analysis. Only the direct inspection of the raw data revealed the source of the heterogeneous treatment group variances. The untestable assumption of a mislabelling followed by removal of the data from channel 16 still led to a straightforward omnibus F-test and ANOVA table.

Great care is also required in the more detailed contrast analysis. While the RTV-only condition was initially thought to provide another control condition, a comparison with the empty control revealed a systematic and substantial difference, with RTV-only showing larger tumor diameters. Two sets of contrasts are then plausible, using different 'baseline' values for estimating the effect of RTV in conjunction with a drug. Both sets of contrasts have straightforward interpretations, but which set is more meaningful depends on the biological question. Note that the important contrast of tumor diameter differences between inhibited and uninhibited conditions compared between the two drugs is independent of this decision.

5.5   Notes and Summary

Notes
Linear contrasts for finding minimum effective doses are discussed in Ruberg (1989), and general design and analysis of dose finding studies in Ruberg (1995a, b).

If and when multiple comparison procedures are required is sometimes a matter of debate. Some people argue that such adjustments are rarely called for in designed experiments (Finney 1988; O'Brien 1983). Further discussion of this topic is provided in Cox (1965), O'Brien (1983), Curran-Everett (2000), and Noble (2009). An authoritative treatment of the issue is Tukey (1991). A book-length treatment is Rupert (2012).

In addition to the weak and strong control family-wise error (Proschan and Brittain 2020), we can also control for other types of error (Lawrence 2019), most prominently the false discovery rate FDR (Benjamini and Hochberg 1995). A relatively recent graphical method allows adapting the Bonferroni–Holm procedure to a specific set of hypotheses and can yield less conservative adjustments (Bretz 2009).

Using R

Estimation of contrasts in R is discussed in Sect. 5.2.4. A very convenient option for applying multiple comparisons procedures is to use the emmeans package and follow the same strategy as before: estimate the model parameters using aov() and estimate the group means using emmeans(). We can then use the contrast() function with an adjust= argument to choose a multiple correction procedure to adjust p-values and confidence intervals of contrasts. This function also has several frequently used sets of contrasts built-in, such as method="pairwise" for generating all pair-wise contrasts or method="trt.vs.ctrl1" and method="trt.vs.ctrlk" for generating contrasts comparing all treatments to the first, respectively, last, level of the treatment factor. For estimated marginal means em, and either the corresponding built-in contrasts or our manually defined set of contrasts ourContrasts, we access the five procedures as

```
contrast(em, method = ourContrasts, adjust = "bonferroni")
contrast(em, method = ourContrasts, adjust = "holm")
contrast(em, method = "pairwise", adjust = "tukey")
contrast(em, method = "trt.vs.ctrl1", adjust = "dunnett")
contrast(em, method = ourContrasts, adjust = "scheffe")
```

By default, these functions provide the contrast estimates and associated t-tests. We can use the results of contrast() as an input to confint() to get contrast estimates and their adjusted confidence intervals instead.

The package Superpower provides functionality to perform power analysis of contrasts in conjunction with emmeans.

Summary

Linear contrasts are a principled way for defining comparisons between two sets of group means and constructing the corresponding estimators, their confidence intervals, and t-statistics. While an ANOVA omnibus F-test looks for any pattern of deviation between group means, linear contrasts use specific comparisons and are more powerful in detecting the specified deviations. Without much exaggeration, linear contrasts are the main reason for conducting comparative experiments and their definition is an important part of an experimental design.

With more than one hypothesis tested, multiple comparison procedures are often required to adjust for the inflation in false positives. General purpose procedures are easy to use, but sets of contrasts often have more structure that can be exploited to gain more power.

Power analysis for contrasts poses no new problems, but the adjustments by MCPs can only be considered for single-step procedures, because multi-step procedures depend on the observed p-values which are of course unknown at the time of planning the experiment.

References

Abelson, R. P. and D. A. Prentice (1997). "Contrast tests of interaction hypothesis". In: Psychological Methods 2.4, pp. 315–328.

Benjamini, Y. and Y. Hochberg (1995). "Controlling the False Discovery Rate: A Practical and Powerful Approach to Multiple Testing". In: Journal of the Royal Statistical Society. Series B (Methodological) 57.1, pp. 289–300.

Bretz, F. et al. (2009). "A graphical approach to sequentially rejective multiple test procedures". In: Statistics in Medicine 28.4, pp. 586–604.

Cox, D. R. (1965). "A remark on multiple comparison methods". In: Technometrics 7.2, pp. 223–224.

Curran-Everett, D. (2000). "Multiple comparisons: philosophies and illustrations". In: American Journal of Physiology-Regulatory, Integrative and Comparative Physiology 279, R1–R8.

Dunnett, C. W. (1955). "A multiple comparison procedure for comparing several treatments with a control". In: Journal of the American Statistical Association 50.272, pp. 1096–1121.

Finney, D. J. (1988). "Was this in your statistics textbook? III. Design and analysis". In: Experimental Agriculture 24, pp. 421–432.

Lawrence, J. (2019). "Familywise and per-family error rates of multiple comparison procedures". In: Statistics in Medicine 38.19, pp. 1–13.

Lohasz, C. et al. (2020). "Predicting Metabolism-Related Drug-Drug Interactions Using a Micro-physiological Multitissue System". In: Advanced Biosystems 4.11, pp. 2000079.

Noble, W. S. (2009). "How does multiple testing correction work?" In: Nature Biotechnology 27.12, pp. 1135–1137.

O'Brien, P. C. (1983). "The appropriateness of analysis of variance and multiple-comparison procedures". In: Biometrics 39.3, pp. 787–788.

Proschan, M. A. and E. H. Brittain (2020). "A primer on strong vs weak control of familywise error rate". In: Statistics in Medicine 39.9, pp. 1407–1413.

Ruberg, S. J. (1989). "Contrasts for identifying the minimum effective dose". In: Journal of the American Statistical Association 84.407, pp. 816–822.

Ruberg, S. J. (1995a). "Dose response studies I. Some design considerations". In: *Journal of Biopharmaceutical Statistics* 5.1, pp. 1–14.

Ruberg, S. J. (1995b). "Dose response studies II. Analysis and interpretation". In: Journal of Biopharmaceutical Statistics 5.1, pp. 15–42.

Rupert Jr, G. (2012). Simultaneous statistical inference. Springer Science & Business Media.

Scheffé, H. (1959). The Analysis of Variance. John Wiley & Sons, Inc.

Tukey, J. W. (1949a). "Comparing Individual Means in the Analysis of Variance". In: Biometrics 5.2, pp. 99–114.

Tukey, J. W. (1991). "The philosophy of multiple comparisons". In: Statistical Science 6, pp. 100–116.

Chapter 6
Multiple Treatment Factors: Factorial Designs

6.1 Introduction

The treatment design in our drug example contains a single treatment factor, and one of four drugs is administered to each mouse. *Factorial treatment designs* use several treatment factors, and a treatment applied to an experimental unit is then a combination of one level from each factor.

While we analyzed our tumor diameter example as a one-way analysis of variance, we can alternatively interpret the experiment as a factorial design with two treatment factors: **Drug** with three levels, 'none', 'CP', and 'IFF', and **Inhibitor** with two levels, 'none' and 'RTV'. Each of the six previous treatment levels is then a combination of one drug and the absence/presence of RTV.

Factorial designs can be analyzed using a *multi-way analysis of variance*—a relatively straightforward extension from Chap. 4—and linear contrasts. A new phenomenon in these designs is the potential presence of *interactions* between two (or more) treatment factors. The difference between the levels of the first treatment factor then depends on the level of the second treatment factor, and greater care is required in the interpretation.

6.2 Experiment

In Chap. 4, we suspected that the four drugs might show different effects under different diets but ignored this aspect by using the same low-fat diet for all treatment groups. We now consider the diet again and expand our investigation by introducing a second diet with high fat content. We previously found that drugs $D1$ and $D2$ from class A are superior to the two drugs from class B, and therefore concentrate on class A exclusively. In order to establish and quantify the effect of these two drugs compared to 'no effect', we introduce a control group using a placebo treatment with no active substance to provide these comparisons within our experiment.

© Springer Nature Switzerland AG 2021
H.-M. Kaltenbach, *Statistical Design and Analysis of Biological Experiments*,
Statistics for Biology and Health, https://doi.org/10.1007/978-3-030-69641-2_6

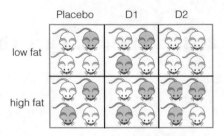

Fig. 6.1 Experiment with two crossed treatment factors with four mice in each treatment combination

Table 6.1 Enzyme levels for combinations of three drugs and two diets, four mice per combination

	Low fat				High fat			
	1	2	3	4	1	2	3	4
Placebo	10.12	8.56	9.34	8.70	10.79	12.01	9.11	10.71
D1	14.32	13.58	14.81	13.98	14.08	13.86	13.94	16.36
D2	13.61	13.24	10.35	13.31	9.84	8.84	12.04	8.58

The proposed experiment is as follows: two drugs, $D1$ and $D2$, and one placebo treatment are given to 8 mice each, and treatment allocation is completely at random. For each drug, 4 mice are randomly selected and receive a low-fat diet, while the other 4 mice receive a high-fat diet. The treatment design then consists of two treatment factors, **Drug** and **Diet**, with three and two levels, respectively. Since each drug is combined with each diet, the two treatment factors are crossed, and the design is balanced because each drug–diet combination is allocated to four mice (Fig. 6.1).

The 24 measured enzyme levels for the six treatment combinations are shown in Table 6.1 and Fig. 6.2A. A useful alternative visualization is an *interaction plot* as shown in Fig. 6.2B, where we show the levels of **Diet** on the horizontal axis, the enzyme levels on the vertical axis, and the levels of **Drug** are represented by point shapes. We also show the average response in each of the six treatment groups and connect those of the same drug by a line, linetype indicating the drug.

Even a cursory glance reveals that both drugs have an effect on the enzyme level, resulting in higher observed enzyme levels compared to the placebo treatment on a low-fat diet. Two drug treatments—$D1$ and placebo—appear to have increased enzyme levels under a high-fat diet, while levels are lower under a high-fat diet for $D2$. This is an indication that the effect of the diet is not the same for the three drug treatments and there is thus an *interaction* between drug and diet.

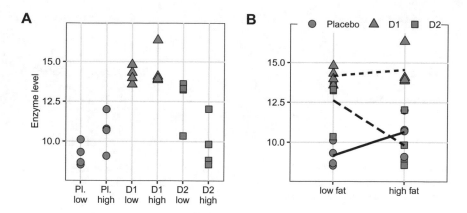

Fig. 6.2 **A** Enzyme levels for combinations of three drugs and two diets, four mice each in a completely randomized design with two-way crossed treatment structure. Drugs are highlighted as point shapes. **B** Same data as interaction plot, with shape indicating drug treatment and lines connecting the average response values for each drug over diets

A crucial point in this experiment is the independent and random assignment of *combinations* of treatment factor levels to each mouse. We have to be careful not to break this independent assignment in the implementation of the experiment. If the experiment requires multiple cages, for example, it is very convenient to feed the same diet to all mice in a cage. This is not a valid implementation of the experimental design, however, because the diet is now randomized on cages and not on mice, and this additional structure is not reflected in the design; we return to this problem when we discuss *split-unit designs* in Chap. 8.

Hasse Diagrams

The treatment structure of this experiment is shown in Fig. 6.3A and is a *factorial treatment design* with two crossed treatment factors **Drug** and **Diet**, which we write next to each other in the treatment structure diagram. Each treatment combination is a level of the new *interaction factor* **Drug:Diet**, nested in both treatment factors.

The unit structure only contains the single unit factor *(Mouse)*, as shown in Fig. 6.3B.

To derive the experiment structure in Fig. 6.3C, we note that each combination of a drug and diet is randomly assigned to a mouse. This makes *(Mouse)* the experimental unit for **Drug** and **Diet** and also for their interaction **Drug:Diet**. We consequently draw an edge from **Drug:Diet** to *(Mouse)*; the edges from **Drug** respectively **Diet** to *(Mouse)* are then 'shortcuts' and are omitted from the diagram.

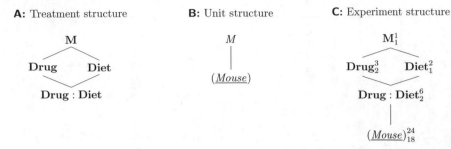

Fig. 6.3 Completely randomized design for determining the effects of two different drugs and placebo combined with two different diets using four mice per combination of drug and diet, and a single response measurement per mouse

6.3 ANOVA for Two Treatment Factors

The analysis of variance framework readily extends to more than one treatment factor. We already find all the new properties of this extension with only two factors, and we can therefore focus on this case for the moment.

6.3.1 Linear Model

For our general discussion, we consider two treatment factors **A** with a levels and **B** with b levels; the interaction factor **A:B** then has $a \cdot b$ levels. We again denote by n the number of experimental units for each treatment combination in a balanced design, and by $N = a \cdot b \cdot n$ the total sample size. For our example, $a = 3$, $b = 2$, $n = 4$, and thus $N = 24$.

Cell Means Model

A simple linear model is the *cell means model*

$$y_{ijk} = \mu_{ij} + e_{ijk} ,$$

in which each cell mean μ_{ij} is the average response of the experimental units receiving level i of the first treatment factor and level j of the second treatment factor, and $e_{ijk} \sim N(0, \sigma_e^2)$ is the residual of the kth replicate for the treatment combination (i, j).

Our example leads to $3 \times 2 = 6$ cell means $\mu_{11}, \ldots, \mu_{32}$; the cell mean μ_{11} is the average response to the placebo treatment under the low-fat diet, and μ_{32} is the average response to the $D2$ drug treatment under the high-fat diet.

The cell means model does not explicitly reflect the factorial treatment design, and we could analyze our experiment as a one-way ANOVA with six treatment levels, as we did in the tumor diameter example in Sect. 5.4.

Parametric Model

The *parametric model*

$$y_{ijk} = \underbrace{\mu_{..}}_{\mu} + \underbrace{(\mu_{i.} - \mu_{..})}_{\alpha_i} + \underbrace{(\mu_{.j} - \mu_{..})}_{\beta_j} + \underbrace{(\mu_{ij} - \mu_{i.} - \mu_{.j} + \mu_{..})}_{(\alpha\beta)_{ij}} + e_{ijk}$$

makes the factorial treatment design explicit. Each cell mean μ_{ij} is decomposed into four contributions: a grand mean μ, the average deviation of cell mean $\mu_{i.}$ for level i of **A** (averaged over all levels of **B**), the average deviation of cell mean $\mu_{.j}$ of **B** (averaged over all levels of **A**), and an interaction for each level of **A:B**.

The model components are shown schematically in Fig. 6.4A for our drug–diet example. Each drug corresponds to a row, and $\mu_{i.}$ is the average response for drug i, where the average is taken over both diets in row i. Each diet corresponds to a column, and $\mu_{.j}$ is the average response for column j (over the three drugs). The interaction is the difference between the additive effect of the row and column differences to the grand mean $\mu_{i.} - \mu_{..}$ and $\mu_{.j} - \mu_{..}$, and the actual cell mean's difference $\mu_{ij} - \mu_{..}$ from the grand mean:

$$\mu_{ij} - \mu_{i.} - \mu_{.j} + \mu_{..} = (\mu_{ij} - \mu_{..}) - \big((\mu_{i.} - \mu_{..}) + (\mu_{.j} - \mu_{..})\big) .$$

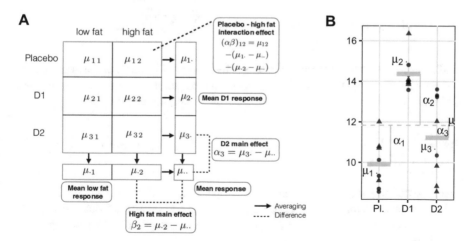

Fig. 6.4 A The 3-by-2 data table with averages per treatment. The main effects are differences between row- or column means and the grand mean. Interaction effects are differences between cell means and the two main effects. **B** Marginal data considered for drug pools data for each drug over diets. The three averages correspond to the row means and point shapes correspond to diet. Dashed line: grand mean; solid gray lines: group means; vertical lines: group mean deviations

The model has five sets of parameters: a parameters α_i and b parameters β_j constitute the *main effect parameters* of factor **A** and **B**, respectively, while the $a \cdot b$ parameters $(\alpha\beta)_{ij}$ are the *interaction effect parameters* of factor **A:B**. They correspond to the factors in the experiment diagram: in our example, the factor **M** is associated with the grand mean μ, while the two treatment factors **Drug** and **Diet** are associated with α_i and β_j, respectively, and the interaction factor **Drug:Diet** with $(\alpha\beta)_{ij}$. The unit factor *(Mouse)* has 24 levels, each reflecting the deviation of a specific mouse, and is represented by the residuals e_{ijk} and their variance σ_e^2. The degrees of freedom are the number of independent estimates for each set of parameters.

6.3.2 Analysis of Variance

The analysis of variance with more than one treatment factor is a direct extension of the one-way ANOVA, and is again based on decomposing the total sum of squares and degrees of freedom. Each treatment factor can be tested individually using a corresponding F-test.

Sums of Squares

For balanced designs, we decompose the sum of squares into one part for each factor in the design. The total sum of squares is

$$SS_{tot} = SS_A + SS_B + SS_{A:B} + SS_{res} = \sum_{i=1}^{a}\sum_{j=1}^{b}\sum_{k=1}^{n}(y_{ijk} - \bar{y}...)^2 \, ,$$

and measures the overall variation between the observed values and the grand mean. We now have two treatment sum of squares, one for factor **A** and one for **B**; they are the squared differences of the group mean from the grand mean:

$$SS_A = \sum_{i=1}^{a}\sum_{j=1}^{b}\sum_{k=1}^{n}(\bar{y}_{i..} - \bar{y}...)^2 = bn\sum_{i=1}^{a}(\bar{y}_{i..} - \bar{y}...)^2 \text{ and } SS_B = an\sum_{j=1}^{b}(\bar{y}_{.j.} - \bar{y}...)^2 \, .$$

Each treatment sum of squares is found by pooling the data over all levels of the other treatment factor. For instance, SS_{drug} in our example results from pooling data over the diet treatment levels, corresponding to a one-way ANOVA situation as in Fig. 6.4B.

The interaction sum of squares for **A:B** is

$$SS_{A:B} = n \sum_{i=1}^{a} \sum_{j=1}^{b} (\bar{y}_{ij.} - \bar{y}_{i..} - \bar{y}_{.j.} + \bar{y}_{...})^2 .$$

If interactions exist, parts of $SS_{tot} - SS_A - SS_B$ are due to a systematic difference between the additive prediction for each group mean and the actual mean of that treatment combination.

The residual sum of squares measures the distances of responses to their corresponding treatment means:

$$SS_{res} = \sum_{i=1}^{a} \sum_{j=1}^{b} \sum_{k=1}^{n} (y_{ijk} - \bar{y}_{ij.})^2 .$$

For our example, we find that $SS_{tot} = 129.44$ decomposes into treatments $SS_{drug} = 83.73$ and $SS_{diet} = 0.59$, interaction $SS_{drug:diet} = 19.78$, and residual $SS_{res} = 25.34$ sums of squares.

Degrees of Freedom

The total degrees of freedom also partition by the corresponding factors, and can be calculated quickly and easily from the Hasse diagram. For the general two-way ANOVA, we have

$$df_{tot} = abn - 1 = df_A + df_B + df_{A:B} + df_{res} = (a-1) + (b-1) + (a-1)(b-1) + ab(n-1).$$

With $a = 3, b = 2, n = 4$ for our example, we find $df_{drug} = 2$, $df_{diet} = 1$, $df_{drug:diet} = 2$, and $df_{res} = 18$, in correspondence with the Hasse diagram.

Mean Squares

Mean squares are found by dividing the sum of squares terms by their degrees of freedom. For example, we can calculate the total mean squares

$$MS_{tot} = \frac{SS_{tot}}{abn - 1} ,$$

which correspond to the variance in the data ignoring treatment groups. Further, $MS_A = SS_A/df_A, MS_B = SS_B/df_B, MS_{A:B} = SS_{A:B}/df_{A:B}$, and $MS_{res} = SS_{res}/df_{res}$. In our example, $MS_{drug} = 41.87, MS_{diet} = 0.59, MS_{drug:diet} = 9.89$, and $MS_{res} = 1.41$.

F-tests

We can perform an omnibus F-test for each factor by comparing its mean squares with the corresponding residual mean squares. We find the correct denominator mean squares from the experiment structure diagram: it corresponds to the closest random factor below the tested treatment factor and is MS_{res} from *(Mouse)* for all three tests in our example.

The two *main effect tests* for **A** and **B** based on the F-statistics

$$F = \frac{MS_A}{MS_{res}} \sim F_{a-1,ab(n-1)} \quad \text{and} \quad F = \frac{MS_B}{MS_{res}} \sim F_{b-1,ab(n-1)}$$

test the two hypotheses

$$H_{0,A} : \mu_{1\cdot} = \cdots = \mu_{a\cdot} \quad \text{and} \quad H_{0,B} : \mu_{\cdot 1} = \cdots = \mu_{\cdot b} \,,$$

respectively. These hypotheses state that the corresponding treatment groups, averaged over the levels of the other factor, have equal means, and the average effect of the corresponding factor is zero. The treatment mean squares have expectations

$$\mathbb{E}(MS_A) = \sigma_e^2 + \frac{nb}{a-1} \sum_{i=1}^{a} \alpha_i^2 \quad \text{and} \quad \mathbb{E}(MS_B) = \sigma_e^2 + \frac{na}{b-1} \sum_{j=1}^{b} \beta_j^2 \,,$$

which both provide an independent estimate of the residual variance σ_e^2 if the corresponding null hypothesis is true.

The *interaction test* for **A:B** is based on the F-statistic

$$F = \frac{MS_{A:B}}{MS_{res}} \sim F_{(a-1)(b-1),ab(n-1)} \,,$$

and the mean squares term has expectation

$$\mathbb{E}(MS_{A:B}) = \sigma_e^2 + \frac{n}{(a-1)(b-1)} \sum_{i=1}^{a} \sum_{j=1}^{b} (\alpha\beta)_{ij}^2 \,.$$

Thus, the corresponding null hypothesis is $H_{0,A:B} : (\alpha\beta)_{ij} = 0$ for all i, j. This is equivalent to

$$H_{0,A:B} : \mu_{ij} = \mu + \alpha_i + \beta_j \text{ for all } i, j \,,$$

stating that each treatment group mean has a contribution from both factors, and these contributions are independent of each other.

ANOVA Table

Using R, these calculations are done easily using `aov()`. We find the model specification from the Hasse diagrams. All fixed factors are in the treatment structure and provide the terms `1+drug+diet+drug:diet` which we abbreviate as `drug*diet`. All random factors are in the unit structure, providing `Error(mouse)`, which we may drop from the specification since *(Mouse)* is the lowest random unit factor. The model `y~drug*diet` yields the ANOVA table

	Df	Sum Sq	Mean Sq	F value	Pr(>F)
Drug	2	83.73	41.87	29.74	1.97e-06
Diet	1	0.59	0.59	0.42	5.26e-01
Drug:Diet	2	19.78	9.89	7.02	5.56e-03
Residuals	18	25.34	1.41		

We find that the sums of squares for the drug effect are large compared to all other contributors and **Drug** is highly significant. The **Diet** main effect does not achieve significance at any reasonable level, and its sums of squares are tiny. However, we find a large and significant **Drug:Diet** interaction and must therefore be very careful in our interpretation of the results, as we discuss next.

6.3.3 *Interpretation of Main Effects*

The *main effect* of a treatment factor is the deviation between group means of its factor levels, averaged over the levels of all other treatment factors.

In our example, the null hypothesis of the drug main effect is

$$H_{0,\text{drug}} : \ \mu_{1.} = \mu_{2.} = \mu_{3.} \ \text{ or equiv. } \ H_{0,\text{drug}} : \ \alpha_1 = \alpha_2 = \alpha_3 = 0 \,,$$

and the corresponding hypothesis for the diet main effect is

$$H_{0,\text{diet}} : \ \mu_{.1} = \mu_{.2} \ \text{ or equiv. } \ H_{0,\text{diet}} : \ \beta_1 = \beta_2 = 0 \,.$$

The first hypothesis asks if there is a difference in the enzyme levels between the three drugs, ignoring the diet treatment by averaging the observations for low-fat and high-fat diets for each drug. The second hypothesis asks if there is a difference between enzyme levels for the two diets, averaged over the three drugs.

From the ANOVA table, we see that the main effect of the drug treatment is highly significant, indicating that regardless of the diet, the drugs affect enzyme levels differently. In contrast, the diet main effect is negligibly small with a large *p*-value, but the raw data in Fig. 6.2A show that while the diet has no influence on the drug levels when averaged over all drug treatments, it has visible effects for *some* drugs: the effect of $D1$ seems to be unaffected by the diet, while a low-fat diet yields higher enzyme levels for $D2$, but lower levels for placebo.

6.3.4   Interpretation of Interactions

The interaction effects quantify how different the effect of one factor is for different levels of another factor. For our example, the large and significant interaction factor shows that while the diet main effect is negligible, the two diets have a very different effect depending on the drug treatment. Simply because the diet main effect is not significant therefore does not mean that the diet effect can be neglected for each drug. The most important rule when analyzing designs with factorial treatment structures is therefore

> **Be careful when interpreting main effects in the presence of large interactions.**

Four illustrative examples of interactions are shown in Fig. 6.5 for two factors **A** and **B** with two levels each. In each panel, the two levels of **A** are shown on the horizontal axis, and a response on the vertical axis. The two lines connect the same levels of factor **B** between levels of **A**.

In the first case (Fig. 6.5A), the two lines are parallel and the difference in response between the low to the high levels of **B** is the same for both levels of **A**, and vice versa. The average responses μ_{ij} in each of the four treatment groups are then fully described by an *additive model*

$$\mu_{ij} = \mu + \alpha_i + \beta_j \, ,$$

with no interaction and $(\alpha\beta)_{ij} = 0$.

In our example, we might compare the responses to placebo and drug $D1$ between diets with almost parallel lines in Fig. 6.2B. The two cell means μ_{11} for placebo and μ_{21} for $D1$ under a low-fat diet are

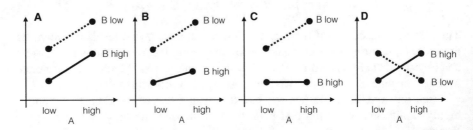

Fig. 6.5 Four stylized scenarios with different interactions. **A** Parallel lines indicate no interaction and an additive model. **B** Small interaction and non-zero main effects. Factor B attenuates effect of factor A. **C** Strongly pronounced interaction. Factor B on high level completely compensates effect of factor A. **D** Complete effect reversal; all information on factor effects is in the interaction

$$\mu_{11} = \mu + \alpha_1 + \beta_1 \quad \text{and} \quad \mu_{21} = \mu + \alpha_2 + \beta_1 ,$$

with difference $\mu_{21} - \mu_{11} = \alpha_2 - \alpha_1$. Under a high-fat diet, the cell means are

$$\mu_{12} = \mu + \alpha_1 + \beta_2 \quad \text{and} \quad \mu_{22} = \mu + \alpha_2 + \beta_2 ,$$

and their difference is also $\mu_{22} - \mu_{12} = \alpha_2 - \alpha_1$.

In other words, the difference between the two drugs is independent of the diet and conversely, the difference between the two diets is independent of the drug administered. We can then interpret the drug effect without regard for the diet and speak of a 'drug effect' or conversely interpret the diet effect independently of the drug and speak of a 'diet effect'.

It might also be that the true situation for placebo and $D1$ is more similar to Fig. 6.5B, a *quantitative interaction*: here, the average response is again always higher if factor **B** is on its low compared to its high level. However, the difference between these two levels now depends on the level of factor **A** and is larger if **A** is on its high compared to its low level. Then, the effects of treatments **A** and **B** are not simply additive, and the interaction term

$$(\alpha\beta)_{ij} = \mu_{ij} - (\mu + \alpha_i + \beta_j)$$

quantifies exactly how much the observed cell mean μ_{ij} differs from the value predicted by the additive model $\mu + \alpha_i + \beta_j$.

For our example, the responses to placebo and drug $D1$ under a low-fat diet are then

$$\mu_{11} = \mu + \alpha_1 + \beta_1 + (\alpha\beta)_{11} \quad \text{and} \quad \mu_{21} = \mu + \alpha_2 + \beta_1 + (\alpha\beta)_{21} ,$$

and thus the difference is $\mu_{21} - \mu_{11} = (\alpha_2 - \alpha_1) + ((\alpha\beta)_{21} - (\alpha\beta)_{11})$, while for the high-fat diet, we find a difference between placebo and $D1$ of $\mu_{22} - \mu_{12} = (\alpha_2 - \alpha_1) + ((\alpha\beta)_{22} - (\alpha\beta)_{12})$. If at least one of the parameters $(\alpha\beta)_{ij}$ is not zero, then the difference between $D1$ and placebo depends on the diet. The interaction is the *difference of differences*

$$\underbrace{(\mu_{21} - \mu_{11})}_{\text{low fat}} - \underbrace{(\mu_{22} - \mu_{12})}_{\text{high fat}} = ((\alpha\beta)_{21} - (\alpha\beta)_{11}) - ((\alpha\beta)_{22} - (\alpha\beta)_{12})$$

and is fully described by the four interaction parameters corresponding to the four means involved.

It is then misleading to speak of a 'drug effect', since this effect is diet-dependent. In our case of a quantitative interaction, we find that a placebo treatment always gives lower enzyme levels, independent of the diet. The diet only *modulates* this effect, and we can achieve a greater increase in enzyme levels for $D1$ over placebo if we additionally use a low-fat diet.

The interaction becomes more pronounced the more 'non-parallel' the lines become. The interaction in Fig. 6.5C is still quantitative, and a low level of **B** always yields the higher response, but the modulation by **A** is much stronger than before. In particular, **A** has no effect under a high level of **B** and using either a low or high level yields the same average response, while under a low level of **B**, using a high level for **A** yields even higher responses.

In our example, there is an additional quantitative interaction between diet and $D1/D2$, with $D1$ consistently giving higher enzyme levels than $D2$ under both diets. However, if we would only consider a low-fat diet, then the small difference between responses to $D1$ and $D2$ might make us prefer $D2$ if it is much cheaper or has a better safety profile, for example, while $D2$ is no option under a high-fat diet.

Interactions can also be *qualitative*, in which case we might see an *effect reversal*. Now, one level of **B** is no longer universally superior under all levels of **A**, as shown in Fig. 6.5D for an extreme case. If **A** is on the low level, then higher responses are gained if **B** is on the low level, while for **A** on its high level, the high level of **B** gives higher responses.

We find this situation in our example when comparing the placebo treatment and $D2$: under a low-fat diet, $D2$ shows higher enzyme levels, while enzyme levels are similar or even lower for $D2$ than placebo under a high-fat diet. In other words, $D1$ is always superior to placebo and we can universally recommend its use, but we should only use $D2$ for a low-fat diet since it seems to have little to no (or even negative) effect under a high-fat diet.

6.3.5 Effect Sizes

We measure the effect size of each factor using its f^2 value. For our generic two-factor design with factors **A** and **B** with a and b levels, respectively, the effect sizes are

$$f_A^2 = \frac{1}{a\sigma_e^2} \sum_{i=1}^{a} \alpha_i^2, \quad f_B^2 = \frac{1}{b\sigma_e^2} \sum_{j=1}^{b} \beta_j^2, \quad f_{A:B}^2 = \frac{1}{ab\sigma_e^2} \sum_{i=1}^{a} \sum_{j=1}^{b} (\alpha\beta)_{ij}^2,$$

which in our example are $f_{\text{drug}}^2 = 0.43$ (huge), $f_{\text{diet}}^2 = 0.02$ (tiny), and $f_{\text{drug:diet}}^2 = 0.3$ (medium).

Alternatively, we can measure the effect of factor X by its variance explained $\eta_X^2 = SS_X/SS_{\text{tot}}$, which in our example yields $\eta_{\text{drug}}^2 = SS_{\text{drug}}/SS_{\text{tot}} = 65\%$ for the drug main effect, $\eta_{\text{diet}}^2 = 0.45\%$ for the diet main effect, and $\eta_{\text{drug:diet}}^2 = 15\%$ for the interaction. This effect size measure has two caveats for multi-way ANOVAs, however. First, the total sum of squares contains all those parts of the variation 'explained' by the other treatment factors. The magnitude of any η_X^2 therefore depends on the number of other treatment factors, and on the proportion of variation explained by them. The second caveat is a consequence of this: the explained variance η_X^2 does not have a simple relationship with the effect size f_X^2, which compares the variation due to the factor alone to the residual variance and excludes all effects due to other treatment factors.

We resolve both issues by the *partial-η^2* effect size measure, which is the fraction of variation explained by each factor over the variation that remains to be explained after accounting for all other treatment factors:

$$\eta_{p,A}^2 = \frac{SS_A}{SS_A + SS_{res}}, \quad \eta_{p,B}^2 = \frac{SS_B}{SS_B + SS_{res}}, \quad \eta_{p,A:B}^2 = \frac{SS_{A:B}}{SS_{A:B} + SS_{res}}.$$

Note that $\eta_{p,X}^2 = \eta_X^2$ for a one-way ANOVA. The partial-η^2 are not a partition of the variation into single-factor contributions and therefore do not sum to one, but they have a direct relation to f^2:

$$f_X^2 = \frac{\eta_{p,X}^2}{1 - \eta_{p,X}^2}$$

for any treatment factor X. In our example, we find $\eta_{p,\,drug}^2 = 77\%$, $\eta_{p,\,diet}^2 = 2\%$, and $\eta_{p,\,drug:diet}^2 = 44\%$.

6.3.6 Model Reduction and Marginality Principle

In the one-way ANOVA, the omnibus F-test was of limited use for addressing our scientific questions. With more factors in the experiment design, however, a non-significant F-test and a small effect size would allow us to remove the corresponding factor from the model. This is known as *model reduction*, and the reduced model is often easier to analyze and interpret.

In order to arrive at an interpretable reduced model, we need to take account of the *principle of marginality* for model reduction (Nelder 1994):

> *If we remove a factor from the model, then we also need to remove all interaction factors involving this factor.*

For a two-way ANOVA, the marginality principle implies that only the interaction factor can be removed in the first step. We then re-estimate the resulting additive model and might further remove one or both main effects factors if their effects are small and non-significant. If the interaction factor cannot be removed, then the two-way ANOVA model cannot be reduced.

Indeed, the interpretation of the model would suffer severely if we removed **A** but kept **A:B**, and this model would then describe **B** as *nested* in **A**, in contrast to the actual design which has **A** and **B** crossed. The resulting model is then of the form

$$y_{ijk} = \mu + \alpha_i + (\alpha\beta)_{ij} + e_{ijk}$$

A: Experiment structure   **B:** Experiment structure
(full model)   (incorrectly reduced model)

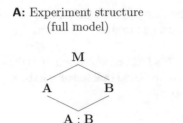   

Fig. 6.6 **A** Experiment structure diagram for a full two-way model with interaction. Factors A and B are crossed. **B** Experiment structure resulting from removing one treatment factor but not its interaction, violating the marginality principle. Factor B is now nested in A contrary to the actual design

and has three sets of parameters: the grand mean, one deviation for each level of **A**, and additionally *different and independent* deviations for the levels of **B** *within* each level of **A**. We see this consequence of violating the marginality principle by comparing the full model diagram in Fig. 6.6A to the diagram of the reduced model in Fig. 6.6B and recalling that two nested factors correspond to a main effect and an interaction effect since A/B=A+A:B.

In our example, we find that **Diet** has a non-significant p-value and small effect size, which suggests that it does not have an appreciable main effect when averaged over drugs. However, **Diet** is involved in the significant interaction factor **Drug:Diet** with large effect size, and the marginality principle prohibits removing **Diet** from the model. Indeed, we have already seen that the effects of the three drugs are appreciably modified by the diet, so we cannot ignore the diet in our analysis.

6.3.7   Estimated Marginal Means

Recall that the estimated marginal means give the average response in a treatment group calculated from a linear model. To solidify our understanding of estimated marginal means, we briefly consider their calculation and interpretation for our example. This also further exemplifies the interpretation of interactions and the consequences of model reduction.

Our example has six possible cell means, one for each combination of a drug with a diet. We consider five linear models for this factorial design with two crossed treatment factors: the full model with interaction, the additive model, two models with a single treatment factor, and a model with no treatment factors. Each of these models

Table 6.2 Empirical means (Data) and predicted means for each cell for full model with interaction (Full), additive model without interactions (Additive), one-way model containing only main effect for drug (Drug only), respectively diet (Diet only), and trivial model with no treatment factors (Average)

Drug	Diet	Data	Full	Additive	Drug only	Diet only	Average
Placebo	Low fat	9.18	9.18	10.07	9,92	11.99	11.84
D1	Low fat	14.17	14.17	14.52	14.37	11.99	11.84
D2	Low fat	12.63	12.63	11.38	11.23	11.99	11.84
Placebo	High fat	10.65	10.65	9.76	9.92	11.68	11.84
D1	High fat	14.56	14.56	14.21	14.37	11.68	11.84
D2	High fat	9.83	9.83	11.07	11.23	11.68	11.84

gives rise to different estimated marginal means for the six treatment combinations, and their predicted cell means are shown in Table 6.2.

The **Full** model is $\mu_{ij} = \mu + \alpha_i + \beta_j + (\alpha\beta)_{ij}$ and has specification y~1+drug+diet+drug:diet. It provides six different and independent cell means, which are identical to the empirical averages (**Data**) for each cell.

The **Additive** model is $\mu_{ij} = \mu + \alpha_i + \beta_j$ and is specified by y~1+drug+diet. It does not have an interaction term and the two treatment effects are hence additive, such that the difference between $D1$ and placebo, for example, is $10.07 - 14.52 = -4.45$ for a low-fat diet and $9.76 - 14.21 = -4.45$ for a high-fat diet. In other words, differences between drugs are independent of the diet and vice versa.

The two models **Drug only** and **Diet only** are $\mu_{ij} = \mu + \alpha_i$ (specified as y~1+drug) and $\mu_{ij} = \mu + \beta_j$ (specified as y~1+diet) and completely ignore the respective other treatment. The first gives identical estimated marginal means for all conditions under the same drug and thus predicts three different cell means, while the second gives identical estimated marginal means for all conditions under the same diet, predicting two different cell means. Finally, the **Average** model $\mu_{ij} = \mu$ is specified as y~1 and describes the data by a single common mean, resulting in identical predictions for the six cell means.

Hence, the model used for describing the experimental data determines the estimated marginal means, and therefore the contrast estimates. In particular, nonsense will arise if we use an additive model such as drug+diet for defining the estimated marginal means, and then try to estimate an interaction contrast (cf. Sect. 6.6.2) using these means. The contrast is necessarily zero, and no standard error or t-statistic can be calculated.

6.3.8 A Real-Life Example—Drug Metabolization Continued

In Sect. 5.4, we looked into a real-life example and examined the effect of two anticancer drugs (IFF and CP) and an inhibitor (RTV) on tumor growth. We analyzed the data using a one-way ANOVA based on six treatment groups and concentrated on relevant contrasts. There is, however, more structure in the treatment design, since we are actually looking at combinations of an anticancer drug and an inhibitor. To account for this structure, we reformulate the analysis to a more natural two-way ANOVA with two treatment factors: **Drug** with the three levels *none*, *CP*, and *IFF*, and **Inhibitor** with two levels *none* and *RTV*. Each drug is used with and without an inhibitor, leading to a fully-crossed factorial treatment design with the two factors and their interaction **Drug:Inhibitor**. We already used the two conditions (*none,none*) and (*none,RTV*) as our control conditions in the contrast analysis. The model specification is `Diameter~Drug*Inhibitor` and leads to the ANOVA table

	Df	Sum Sq	Mean Sq	F value	Pr(>F)
Drug	2	495985.17	247992.58	235.05	4.11e-29
Inhibitor	1	238533.61	238533.61	226.09	5.13e-22
Drug:Inhibitor	2	80686.12	40343.06	38.24	1.96e-11
Residuals	60	63303.4	1055.06		

We note that the previous sum of squares for **Condition** was 815204.90 in our one-way analysis and now exactly partitions into the contributions of our three treatment factors. The previous effect size of 93% is also partitioned into $\eta^2_{\text{Drug}} = 56\%$, $\eta^2_{\text{Inhibitor}} = 27\%$, and $\eta^2_{\text{Drug:Inhibitor}} = 9\%$. In other words, the differences between conditions are largely caused by differences between the anticancer drugs, but the presence of the inhibitor also explains more than one-quarter of the variation. The interaction is highly significant and cannot be ignored, but its effect is much smaller compared to the two main effects. All of this is of course in complete agreement with our previous contrast analysis.

6.4   More on Factorial Designs

We briefly discuss the general advantages of factorial treatment designs and extend our discussion to more than two treatment factors.

6.4.1   Advantage of Factorial Treatment Design

Before the advantages of simultaneously using multiple crossed treatment factors in factorial designs were fully understood by practitioners, so-called *one variable at a time (OVAT)* designs were commonly used for studying the effect of multiple treatment factors. Here, only a single treatment factor was considered in each experiment, while all other treatment factors were kept at a fixed level to achieve maximally controlled conditions for studying each factor. These designs do not allow the estimation of interactions and can therefore lead to incorrect interpretation if interactions are present. In contrast, factorial designs reveal interaction effects if they are present and provide the basis for correct interpretation of the experimental effects. The mistaken logic of OVAT designs was clearly recognized by R. A. Fisher in the 1920s:

> *No aphorism is more frequently repeated in connection with field trials, than that we must ask Nature few questions, or, ideally, one question, at a time. The writer is convinced that this view is wholly mistaken. Nature, he suggests, will best respond to a logical and carefully thought out questionnaire; indeed, if we ask her a single question, she will often refuse to answer until some other topic has been discussed.* (Fisher 1926)

Slightly less obvious is the fact that factorial designs also require smaller sample sizes than OVAT designs, even if no interactions are present between the treatment factors. For example, we can estimate the main effects of two factors **A** and **B** with two levels each using two OVAT designs with $2n$ experimental units per experiment. Each main effect is then estimated with $2n - 2$ residual degrees of freedom. We can use the same experimental resources for a 2×2-factorial design with $4n$ experimental units. If the interaction **A:B** is large, then only this factorial design will allow correct inferences. If the interaction is negligible, then we have two estimates of the **A** main effect: first for **B** on the low level by contrasting the n observations for **A** on a low level with those n for **A** on a high level, and second for the $n + n$ observations for **B** on a high level. The **A** main effect estimate is then the average of these two, based on all $4n$ observations; this is sometimes called *hidden replication*. The inference on **A** is also more general, since we observe the **A** effect under two conditions for **B**. The same argument holds for estimating the **B** main effect. Since we need to estimate one grand mean and one independent group mean each for **A** and **B**, the residual degrees of freedom are $4n - 3$.

6.4.2 One Observation per Cell

Constraints on resources or complicated logistics for the experiment sometimes do not permit replication within each experimental condition, and we then have a single observation for each treatment combination. The Hasse diagram in Fig. 6.7A shows the resulting experiment structure for two generic factors **A** and **B** with a and b levels, respectively, and one observation per combination. We do not have degrees of freedom for estimating the residual variance, and we can neither estimate confidence intervals for establishing the precision of estimates nor perform hypothesis tests. If large interactions are suspected, then this design is fundamentally flawed.

Removing Interactions

The experiment can be successfully analyzed using an additive model $\mu_{ij} = \mu + \alpha_i + \beta_j$ if interactions are small enough to be ignored, such that $(\alpha\beta)_{ij} \approx 0$. We can

A: Full model   **B:** Additive model

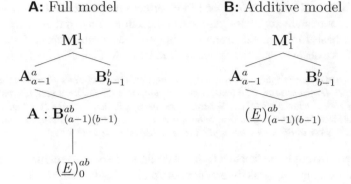

Fig. 6.7 **A** Experiment structure for $a \times b$ two-way factorial with single observation per cell; no residual degrees of freedom left. **B** Same experiment based on additive model removes interaction factor and frees degrees of freedom for estimating residual variance

then remove the interaction factor from the full model, which leads to new residuals $(\alpha\beta)_{ij} + e_{ij} \approx e_{ij}$ and 'frees up' $(a-1)(b-1)$ degrees of freedom for estimating their residual variance. This strategy corresponds to merging the **A:B** interaction factor with the experimental unit factor (E), resulting in the experimental structure in Fig. 6.7B.

Tukey's Method for Testing Interactions

The crucial assumption of no interaction is sometimes justified on subject-matter grounds or from previous experience. If a non-replicated factorial design is run without a clear justification, we can use an ingenious procedure by Tukey to test for an interaction (Tukey 1949b). It applies if at least one of the two treatment factors has three levels or more. Instead of the full model with interaction, Tukey proposed to use a more restricted version of the interaction: a multiple of the product of the two main effects. The corresponding model is

$$y_{ij} = \mu + \alpha_i + \beta_j + \tau \cdot \alpha_i \cdot \beta_j + e_{ij} \, ,$$

an additive model based solely on the main effect parameters, augmented by a special interaction term that introduces a single additional parameter τ and requires only one rather than $(a-1)(b-1)$ degrees of freedom. If the hypothesis $H_0 : \tau = 0$ of no interaction is true, then the test statistic

$$F = \frac{\mathrm{SS}_\tau/1}{\mathrm{MS}_{\mathrm{res}}} \sim F_{1,ab-(a+b)}$$

has an F-distribution with one numerator and $ab - (a + b)$ denominator degrees of freedom.

Here, the 'interaction' sum of squares is calculated as

$$\text{SS}_\tau = \frac{\left(\sum_{i=1}^{a} \sum_{j=1}^{b} y_{ij} \cdot (\bar{y}_{i\cdot} - \bar{y}_{\cdot\cdot}) \cdot (\bar{y}_{\cdot j} - \bar{y}_{\cdot\cdot}) \right)^2}{\left(\sum_{i=1}^{a} (\bar{y}_{i\cdot} - \bar{y}_{\cdot\cdot})^2 \right) \cdot \left(\sum_{j=1}^{b} (\bar{y}_{\cdot j} - \bar{y}_{\cdot\cdot})^2 \right)}$$

and has one degree of freedom.

Lenth's Method for Analyzing Unreplicated Factorials

An alternative method for analyzing a non-replicated factorial design was proposed in Lenth (1989). It crucially depends on the assumption that most effects are small and negligible, and only a fraction of all effects is *active* and therefore relevant for the analysis and interpretation.

The main idea is then to calculate a robust estimate of the standard error based on the assumption that most deviations are random and not caused by treatment effects. This error estimate is then used to discard factors whose effects are sufficiently small to be negligible.

Specifically, we denote the estimated average difference between low and high levels of the jth factor by c_j and estimate the standard error as 1.5 times the median of absolute effect estimates:

$$s_0 = 1.5 \cdot \text{median}_j |c_j| \ .$$

If no effect were active, then s_0 would already provide an approximate estimate of the standard error. If some effects are active, they inflate the estimate by an unknown amount. We therefore restrict our estimation to those effects that are 'small enough' and do not exceed 2.5 times the current standard error estimate. The *pseudo-standard error* is then

$$\text{PSE} = 1.5 \cdot \text{median}_{|c_j| < 2.5 \cdot s_0} |c_j| \ .$$

The *margin of error (ME)* (an upper limit of a confidence interval) is then

$$\text{ME} = t_{0.975, d} \cdot \text{PSE} \ ,$$

and Lenth proposes to use $d = m/3$ as the degrees of freedom, where m is the number of effects in the model. This limit is corrected for multiple comparisons by adjusting the confidence limit from $\alpha = 0.975$ to $\gamma = (1 + 0.95^{1/m})/2$. The resulting *simultaneous margin of error (SME)* is then

$$\text{SME} = t_{\gamma, d} \cdot \text{PSE} \ .$$

Table 6.3 Experimental design and isatin yield of Davies' experiment

S	t	A	T	Yield	S	t	A	T	Yield
−1	−1	−1	−1	0.08	−1	−1	−1	+1	0.79
+1	−1	−1	−1	0.04	+1	−1	−1	+1	0.68
−1	+1	−1	−1	0.53	−1	+1	−1	+1	0.73
+1	+1	−1	−1	0.43	+1	+1	−1	+1	0.08
−1	−1	+1	−1	0.31	−1	−1	+1	+1	0.77
+1	−1	+1	−1	0.09	+1	−1	+1	+1	0.38
−1	+1	+1	−1	0.12	−1	+1	+1	+1	0.49
+1	+1	+1	−1	0.36	+1	+1	+1	+1	0.23

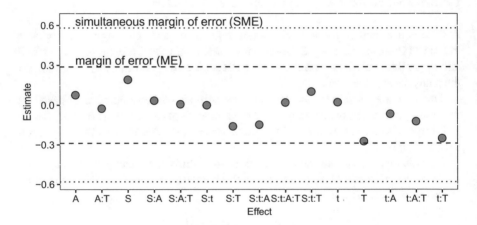

Fig. 6.8 Analysis of active effects in unreplicated 2^4-factorial with Lenth's method

Factors with effects exceeding SME in either direction are considered active, those between the ME limits are inactive, and those between ME and SME have unclear status. We therefore choose those factors that exceed SME as our safe choice, and might include those exceeding ME as well for subsequent experimentation.

In his paper, Lenth discusses a 2^4-factorial experiment, where the effect of acid strength (S), time (t), amount of acid (A), and temperature (T) on the yield of isatin is studied (Davies 1954). The experiment design and the resulting yield are shown in Table 6.3.

The results are shown in Fig. 6.8. No factor seems to be active, with temperature, acid strength, and the interaction of temperature and time coming closest. Note that the marginality principle requires that if we keep the temperature-by-time interaction, we must also keep the two main effects temperature and time in the model, regardless of their size or significance.

6.4.3   Higher-Order Factorials

We can construct factorial designs for any number of factors with any number of levels, but the number of treatment combinations increases exponentially with the number of factors, and the more levels we consider per factor, the more combinations we get. For example, introducing the vendor as a third factor with two levels into our drug–diet example already yields $3 \cdot 2 \cdot 2 = 12$ combinations, and introducing a fourth three-level factor increases this number further to $12 \cdot 3 = 36$ combinations. In addition, both extensions yield higher-order interactions between three and four factors.

A Model with Three Treatment Factors

We illustrate higher-order factorials using a *three-way factorial* with three treatment factors **Drug** with three levels, **Diet** (two levels), and **Vendor** (two levels). The experiment design is shown in Fig. 6.9A with two replicates per treatment. The corresponding linear model is

$$y_{ijkl} = \mu_{ijk} + e_{ijkl} = \mu + \alpha_i + \beta_j + \gamma_k + (\alpha\beta)_{ij} + (\alpha\gamma)_{ik} + (\beta\gamma)_{jk} + (\alpha\beta\gamma)_{ijk} + e_{ijkl} \ .$$

This model has three main effects, three two-way interactions, and one three-way interaction. A large and significant three-way interaction $(\alpha\beta\gamma)_{ijk}$ arises, for example, if the drug–diet interaction that we observed for $D2$ compared to placebo under both diets only occurs for the kit of vendor A, but not for the kit of vendor B (or with different magnitude). In other words, the two-way interaction itself depends on the level of the third factor, and we are now dealing with a difference of a 'difference of differences'.

In this particular example, we might consider reducing the model by removing all interactions involving **Vendor**, as there seems no plausible reason why the kit used for preparing the samples should interact with drug or diet. The resulting diagram is Fig. 6.9B; this model reduction adheres to the marginality principle.

A further example of a three-way analysis of variance inspired by an actual experiment concerns the inflammation reaction in mice and potential drug targets to suppress inflammation. It was already known that a particular pathway gets activated to trigger an inflammation response, and the receptor of that pathway is a known drug target. In addition, there was preliminary evidence that inflammation sometimes triggers even if the pathway is knocked out. A $2 \times 2 \times 2 = 2^3$-factorial experiment was conducted with three treatment factors: (i) genotype: wildtype mice/knock-out mice without the known receptor; (ii) drug: placebo treatment/known drug; (iii) waiting time: 5 h/8 h between inducing inflammation and administering the drug or placebo. Several mice were used for each of the 8 combinations and their protein levels measured using mass spectrometry. The main focus of inference is the genotype-by-drug interaction, which would indicate if an alternate activation exists and how strong it is.

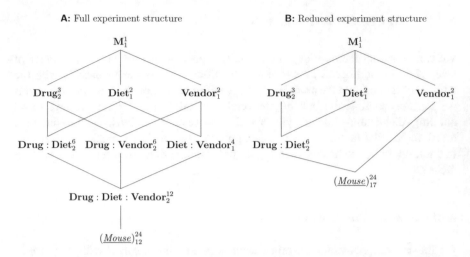

Fig. 6.9 Three-way factorial. **A** Full experiment structure with three two-way and one three-way interaction. **B** Reduced experiment structure assuming negligible interactions with vendor

The three-way interaction then provides additional information about the difference in activation times between the known and a potential unknown pathway. We discuss this example in more detail in Sect. 9.8.6.

Strategies for High-Order Factorials

The implementation and analysis of factorial designs become more challenging as the number of factors and their levels increase. First, the number of parameters and degrees of freedom increase rapidly with the order of the interaction. For example, if three treatment factors are considered with three levels each, then $a = b = c = 3$, yielding one grand mean parameter, $a - 1 = b - 1 = c - 1 = 2$ main effect parameters per factor (6 in total), $(a - 1)(b - 1) = 4$ two-way interaction parameters per interaction (12 in total), and $(a - 1)(b - 1)(c - 1) = 8$ parameters for the three-way interaction, with a total of $1 + 6 + 12 + 8 = 27$ parameters to describe the full model. Of course, this number is identical to the number $3^3 = 27$ of possible treatment combinations (the model is *saturated*), and the experiment already requires 54 experimental units for two replicates.

In addition, the interpretation of an interaction becomes more intricate the higher its order: already a three-way interaction describes a difference of a difference of differences. In practice, interactions of order higher than three are rarely considered in a final model due to difficulties in interpretation, and already three-way interactions are not very common.

If significant higher-order interactions occur with considerable effect sizes, we can try to *stratify* the analysis by one or several factors and provide separate analyses for each of the levels. For our example, with a significant and large three-way interaction, we could stratify by vendor and produce two separate models, one for vendor A and one for vendor B, studying the effect of drug and diet independently for both kits. The advantage of this approach is the easier interpretation, but we are also effectively halving the experiment size, such that information on drug and diet gained from one vendor does not transfer easily into information for the other vendor.

With more than three factors, even higher-order interactions can be considered. Statistical folklore and practical experience suggest that interactions often become less pronounced the higher their order, a heuristic known as *effect sparsity*. Thus, if there are no substantial two-way interactions, we do not expect relevant interactions of order three or higher.

If we have several replicate observations per treatment, then we can estimate the full model with all interactions and use the marginality principle to reduce the model by removing higher-order interactions that are small and non-significant. If only one observation per treatment group is available, a reasonable strategy is to remove the highest-order interaction term from the model in order to free its degrees of freedom. These are then used for estimating the residual variance, which in turn allows us to find confidence intervals and perform hypothesis tests on the remaining factors. The highest-order interaction typically has many degrees of freedom, and we can continue with a model reduction based on F-tests with sufficient power.

Two more strategies are commonly used and are very powerful when applicable. When considering many factors and their low-order interactions, we can use *fractional factorial designs* to deliberately confound some effects and reduce the experiment size. We discuss these designs in detail for the case that each factor has two levels in Chap. 9. The second strategy applies when we are starting out with our experiments, and are unsure which treatment factors to consider for our main experiments. We can then conduct preliminary *screening experiments* to simultaneously investigate a large number of potential treatment factors, and identify those that have sufficient effect on the response. These experiments concentrate on the main effects only and are based on the assumption that only a small fraction of the factors considered will be relevant. This allows experimental designs of small size, and we discuss some options in Sect. 9.7.

6.5 Unbalanced Data

Serious problems can arise in the calculation and interpretation of the analysis of variance for factorial designs with unbalanced data. We briefly review the main problems and some remedies. The defining feature of unbalanced data is succinctly stated as

> *The essence of unbalanced data is that measures of effects of one factor depend upon other factors in the model.* (Littell 2002)

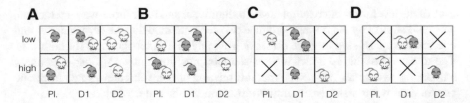

Fig. 6.10 Experiment layout for three drugs under two diets. **A** Unbalanced design with all cells filled but some cells having more data than others. **B, C** Connected some-cells-empty data cause problems in non-additive models. **D** Unconnected some-cells-empty data prohibit estimation of any effects

In particular, the decomposition of the sums of squares is no longer unique for unbalanced data, and the sum of squares for a specific factor depends on the factors preceding it in the model specification. Again, all the problems already appear for a two-factor experiment, and we concentrate on our drug–diet example for illustrating them.

The four data layouts in Fig. 6.10 illustrate different scenarios of unbalanced data: scenario (A) is known as *all-cells-filled data*, where at least one observation is available for each treatment group, but different groups have different numbers of observations. Scenarios (B), (C), and (D) are examples of *some-cells-empty data*, where one or more treatment groups have no observation; the three scenarios are different, as (B) and (C) allow the estimation of main effects and contrasts under an additive model, while no main effects and a very limited number of contrasts can be estimated in (D).

6.5.1 All-Cells-Filled Data

We first consider the case of all-cells-filled data for a $a \times b$ factorial with a rows and b columns. The observations are y_{ijk} with $i = 1 \ldots a$, $j = 1 \ldots b$, and $k = 1 \ldots n_{ij}$, and we can estimate the mean μ_{ij} of row i and column j by taking the average of the corresponding observations y_{ijk} of cell (i, j). The *unweighted row means* are the averages of the cell means over all columns:

$$\mu_{i.} = \frac{1}{b} \sum_{j=1}^{b} \mu_{ij} \, ,$$

and are the population marginal means estimated by the estimated marginal means

$$\hat{\mu}_{i.} = \tilde{y}_{i..} = \frac{1}{b} \sum_{j=1}^{b} \bar{y}_{ij.} \, .$$

The naive *weighted row means*, on the other hand, sum over all observations in each row and divide by the total number of observations in that row; they are

$$\mu_{i\cdot}^w = \sum_{j=1}^{b} \frac{n_{ij}}{n_{i\cdot}} \mu_{ij} \quad \text{estimated as} \quad \hat{\mu}_{i\cdot}^w = \sum_{j=1}^{b} \frac{n_{ij}}{n_{i\cdot}} \bar{y}_{ij\cdot}$$

The unweighted and weighted row means coincide for balanced data with $n_{ij} \equiv n$, but they are usually different for unbalanced data.

We are interested in the row main effect and in testing the hypothesis

$$H_0 : \mu_{1\cdot} = \cdots = \mu_{a\cdot}.$$

of equal (unweighted) row means. In a two-way ANOVA, the corresponding F-test for the main effect of the row factor does *not* test this hypothesis, but rather a hypothesis based on the weighted row means, namely

$$H_0^w : \mu_{1\cdot}^w = \cdots = \mu_{a\cdot}^w.$$

This was not a problem before for balanced data and $\mu_{i\cdot} = \mu_{i\cdot}^w$, but the tested hypothesis for unbalanced data is unlikely to be of any direct interest. The same problems arise for column effects.

We consider an unbalanced version of our drug–diet example as an illustration:

$y_{ij\cdot}(n_{ij})\bar{y}_{ij\cdot}$	Placebo	D1	D2	Row $y_{i\cdot\cdot}(n_{i\cdot})\bar{y}_{i\cdot\cdot}$
low fat	9(1)9	28(2)14	27(2)13.5	64(5)12.8
high fat	23(2)11.5	14(1)14	10(1)10	47(4)11.8
$y_{\cdot j\cdot}(n_{\cdot j})\bar{y}_{\cdot j\cdot}$	32(3)10.7	42(3)14	37(3)12.3	111(9)12.3

Each entry in this table shows the sum $y_{ij\cdot}$ over the responses in the corresponding cell, the number of observations n_{ij} for that cell in parentheses, and the cell average $\bar{y}_{ij\cdot} = y_{ij\cdot}/n_{ij}$. Values in the table margins are the row and column totals and averages and the overall total and average.

For these data, we find an unweighted mean for the first row (the low-fat diet) of $\hat{\mu}_{1\cdot} = (9 + 14 + 13.5)/3 = 12.2$, while the corresponding weighted row mean is $\mu_{1\cdot}^w = (9 + 28 + 27)/(1 + 2 + 2) = 12.8$.

The main problem for a traditional analysis of variance is that for unbalanced data, the sums of squares are not orthogonal and do not decompose uniquely, and there is some part of the overall variation that can be attributed to either of the two factors. For example, the model y~diet*drug would decompose the total sum of squares into

$$SS_{tot} = SS_{\text{Diet adj. mean}} + SS_{\text{Drug adj. Diet}} + SS_{\text{Diet:Drug}} + SS_{res},$$

where $SS_{\text{Diet adj. mean}}$ is the sum of squares attributed to **Diet** after the grand mean has been accounted for, and $SS_{\text{Drug adj. Diet}}$ is the *remaining* sum of squares for **Drug**, once the diet has been accounted for. The traditional ANOVA based on the model y~diet+drug then tests the weighted diet main effect

$$H_0^w : \frac{1\mu_{11} + 2\mu_{12} + 2\mu_{13}}{5} = \frac{2\mu_{21} + 1\mu_{22} + 1\mu_{23}}{4} \; ;$$

this hypothesis depends on the specific number of replicates in each cell and is unlikely to yield any insight into the comparison of drugs and diets in general.

Since the F-statistics consider the treatment mean squares and the residual mean squares, their values and interpretations depend on the order of the factors in the model. For a balanced design, the two decompositions and F-tests are identical, but for unbalanced data, they can deviate.

For our example, we find that the model y~diet*drug yields the ANOVA table

	Df	Sum Sq	Mean Sq	F value	Pr(>F)
Diet	1	2.45	2.45	7.35	7.31e-02
Drug	2	14.44	7.22	21.66	1.65e-02
Diet:Drug	2	12.11	6.06	18.17	2.11e-02
Residuals	3	1	0.33		

On the other hand, the model y~drug*diet decomposes the total sum of squares into

$$SS_{\text{tot}} = SS_{\text{Drug adj. mean}} + SS_{\text{Diet adj. Drug}} + SS_{\text{Diet:Drug}} + SS_{\text{res}} \; ,$$

and produces a larger sum of squares term for **Drug** and a smaller term for **Diet**. In this model, the latter only accounts for the variation after the drug has been accounted for. Even worse, the same analysis of variance based on the additive model y~drug+diet tests yet another and even less interesting (or comprehensible) hypothesis, namely

$$H_0^{w'} : \sum_{j=1}^{b} n_{ij} \cdot \mu_{ij} = \sum_{j=1}^{b} n_{ij} \cdot \mu_{i.}^w \quad \text{for all } i \; .$$

For our example, the full model y~drug*diet yields the ANOVA table

	Df	Sum Sq	Mean Sq	F value	Pr(>F)
Drug	2	16.67	8.33	25	1.35e-02
Diet	1	0.22	0.22	0.67	4.74e-01
Drug:Diet	2	12.11	6.06	18.17	2.11e-02
Residuals	3	1	0.33		

Note that it is only the attribution of variation to the two treatment factors that varies between the two models, while the total sum of squares and the interaction and residual sums of squares are identical. In particular, the sums of squares always add to the total sum of squares. This analysis is known as a *type-I* sum of squares ANOVA, where results depend on the order of factors in the model specification, such that effects of factors introduced later in the model are adjusted for effects of all factors introduced earlier.

In practice, unbalanced data are often analyzed using a *type-II* sum of squares ANOVA based on an additive model without interactions. In this analysis, the treatment sum of squares for *any* factor is calculated after adjusting for all other factors, and the sum of squares do no longer add to the total sum of squares, as part of the variation is accounted for in each adjustment and not attributed to the corresponding factor. The advantage of this analysis is that for the additive model, each F-test corresponds to the respective interesting hypothesis based on the unweighted means, which is independent of the specific number of observations in each cell. In R, we can use the function `Anova()` from package `car` with option `type=2` for this analysis: first, we fit the additive model as `m = aov(y~diet+drug, data=data)` and then call `Anova(m, type=2)` on the fitted model.

For our example, this yields the ANOVA table

	Sum Sq	Df	F value	Pr(>F)
Diet	0.22	1	0.08	7.83e-01
Drug	14.44	2	2.75	1.56e-01
Residuals	13.11	5		

The treatment sums of squares in this table are both adjusted for the respective other factor, and correspond to the second rows in the previous two tables. However, we know that the interaction is important for the interpretation of the data, and the results from this additive model are still misleading.

A similar idea is sometimes used for non-additive models with interactions, known as *type-III* sum of squares. Here, a factor's sum of squares term is calculated after adjusting for all other main effects and all interactions involving the factor. These can be calculated in R using `Anova()` with option `type=3`. Their usefulness is heavily contested by some statisticians, though, since the hypotheses now relate to main effects in the presence of relevant interactions, and are therefore difficult to interpret (some would say useless). This sentiment is expressed in the following scoffing remark:

> *The non-problem that Type III sums of squares tries to solve only arises because it is so simple to do silly things with orthogonal designs. In that case main effect sums of squares are uniquely defined and order independent. Where there is any failure of orthogonality, though, it becomes clear that in testing hypotheses, as with everything else in statistics, it is your responsibility to know clearly what you mean and that the software is faithfully enacting your intentions.* (Venables 2000)

The difficulty of interpreting ANOVA results in the presence of interactions is of course not specific to unbalanced data; the problem of defining suitable main effect tests is just more prominent here and is very well hidden for balanced data. If large interactions are found, we need to resort to more specific analyses in order to make sense of the data. This is most easily done by using estimated marginal means and contrasts between them.

6.5.2 Some-Cells-Empty Data

Standard analyses may or may not apply if some cells are empty, depending on the particular pattern of empty cells. Several situations of some-cells-empty data are shown in Fig. 6.10.

In Fig. 6.10B, only cell $(1, 3)$ is missing and both row and column main effects can be estimated if we assume an additive model. This model allows the prediction of the average response in the missing cell, since the difference $\mu_{22} - \mu_{23}$ is then identical to $\mu_{12} - \mu_{13}$, and we can therefore estimate the missing cell mean by $\mu_{13} = (\mu_{22} - \mu_{23}) - \mu_{12}$; this is an estimated marginal mean for μ_{13}. We can also estimate an interaction involving the first two columns and rows using our previous methods when ignoring the third level of the column factor.

In Fig. 6.10C, several cells are missing, but the remaining cells are still *connected*: by rearranging rows and columns, we can draw a connection from any filled cell to any other filled cell using only vertical and horizontal lines. These data thus provide the information needed to estimate an additive model, and we use the resulting model to estimate the marginal means of the empty cells. In contrast to scenario (B), however, we cannot find a filled 2×2 sub-table and are unable to estimate any interaction.

In Fig. 6.10D, the non-empty cells are not connected, and we cannot predict the empty cell means based on observed data. Neither main nor interaction effects can be estimated for the non-connected part of the table.

Each scenario poses severe problems when an additive model is not adequate and interactions have to be considered. Estimated marginal means can then not be used to compensate empty cells, and the analysis of models with interactions based on some-cells-empty data is necessarily restricted to subsets of rows and columns of the data for which all cells are filled. Similarly, contrast estimates can only be formed if the contrast does not involve empty cells. Such analyses are therefore highly specific to the situation at hand and defy simple recipes.

6.6 Contrast Analysis

Returning to balanced data, linear contrasts allow a more targeted analysis of treatment effects just like in the one-way ANOVA. As before, we define contrasts on the treatment group means μ_{ij}, and we estimate these using the estimated marginal means based on an appropriate model.

6.6.1 Main Effect Contrasts

In our example, the F-test for the interaction factor indicates that the diet affects the drugs differently, and we might use the following contrasts to compare each drug between the diets:

$$\Psi(\mathbf{w}_1) = \mu_{11} - \mu_{12}, \quad \Psi(\mathbf{w}_2) = \mu_{21} - \mu_{22}, \quad \text{and} \quad \Psi(\mathbf{w}_3) = \mu_{31} - \mu_{32},$$

where $\mathbf{w}_1 = (1, 0, 0, -1, 0, 0)$ is the weight vector for comparing the placebo effect between low fat and high fat, and $\mathbf{w}_2 = (0, 1, 0, 0, -1, 0)$ and $\mathbf{w}_3 = (0, 0, 1, 0, 0, -1)$ are the corresponding weight vectors for $D1$ and $D2$.

The results in Table 6.4 give quantitative confirmation of our earlier suspicion: both placebo and $D1$ show lower responses under the low-fat diet compared to high-fat diet, but the resulting difference is explainable by sampling variation and cannot be distinguished from zero with the given data. In contrast, the response for $D2$ is substantially and significantly higher under a low-fat diet compared to a high-fat diet. Confidence intervals are fairly wide due to the small sample size, and the experiment is likely underpowered.

Each contrast's standard error is based on 18 degrees of freedom, and the residual variance estimate is based on all data. The three contrasts are orthogonal, but there are two more orthogonal contrasts between the six cell means in Table 6.5, which we consider next.

6.6.2 Interaction Contrasts

Similar to our previous considerations for the tumor diameter example in Sect. 5.4, we have to ask whether main effect contrasts for the two active drugs $D1$ and $D2$ provide the correct comparisons. If a substantial difference in enzyme levels exists between the diets in the placebo group, then we might rather want a contrast that compares the difference in enzyme levels between diets for $D1$ (respectively $D2$) to the corresponding difference in the placebo group. This contrast is then a difference of differences and thus an interaction contrast.

Table 6.4 Contrast estimates and 95%-confidence intervals for comparing individual drugs under low- and high-fat diets

Contrast	Estimate	se	df	LCL	UCL
Placebo low–Placebo high	−1.48	0.84	18	−3.24	0.29
D1 low–D1 high	−0.39	0.84	18	−2.15	1.37
D2 low–D2 high	2.80	0.84	18	1.04	4.57

Table 6.5 Estimated cell means for three drugs and two diets

Drug	Diet	Mean	Estimate	se	df	LCL	UCL
Placebo	Low fat	μ_{11}	9.18	0.59	18	7.93	10.42
D1	Low fat	μ_{21}	14.17	0.59	18	12.93	15.42
D2	Low fat	μ_{31}	12.63	0.59	18	11.38	13.87
Placebo	High fat	μ_{12}	10.65	0.59	18	9.41	11.90
D1	High fat	μ_{22}	14.56	0.59	18	13.31	15.81
D2	High fat	μ_{32}	9.83	0.59	18	8.58	11.07

For $D1$, we calculate the difference in enzyme levels between the two diets under $D1$ to the difference under placebo with the contrast

$$\Psi(\mathbf{w}_4) = (\mu_{21} - \mu_{22}) - (\mu_{11} - \mu_{12}) .$$

This contrast is equivalent to first 'adjusting' each enzyme level under $D1$ by that of the placebo control group, and then determining the difference between the two adjusted values. The corresponding contrast is

$$\Psi(\mathbf{w}_5) = (\mu_{21} - \mu_{11}) - (\mu_{22} - \mu_{12}) ,$$

which is identical to $\Psi(\mathbf{w}_4)$ after some rearranging of terms. A corresponding contrast for $D2$ is

$$\Psi(\mathbf{w}_6) = (\mu_{31} - \mu_{32}) - (\mu_{11} - \mu_{12}) .$$

The estimates and 95%-confidence intervals are shown in Table 6.6. Not unexpected, we find that the change in response for $D1$ is larger than the change in the placebo group, but not significantly so. We already know that $D1$ enzyme levels are higher for both diets from the main effect contrasts, so this new result means that the difference between placebo and $D1$ is the same for both diets. In other words, the response to $D1$ changes from low fat to high fat by the same amount as does the response to placebo. The two lines in Fig. 6.2B should then be roughly parallel (not considering sample variation), as indeed they are.

In contrast, the change for $D2$ is substantially and significantly larger than that for the placebo group. This is also in accordance with Fig. 6.2B: while the response to placebo is increasing slightly from low fat to high fat, the response to $D2$ does the

Table 6.6 Interaction contrast estimates and Bonferroni-adjusted 95%-confidence intervals

Contrast	Estimate	se	df	LCL	UCL
(D1 low–D1 high)–(Pl. low–Pl. high)	1.09	1.19	18	−1.81	3.99
(D2 low–D2 high)–(Pl. low–Pl. high)	4.28	1.19	18	1.38	7.18

Table 6.7 Orthogonal contrast estimates and Bonferroni-adjusted 95%-confidence intervals

Contrast	Estimate	se	df	LCL	UCL
(D1 low–D1 high)–(D2 low–D2 high)	3.19	1.19	18	0.29	6.09
(Pl. low–Pl. high)–avg(drug low–drug high)	−4.62	1.03	18	−7.13	−2.10

opposite and decreases substantially from low fat to high fat. This is reflected by the significant large positive contrast, which indicates that the change in $D2$ between the two diets follows another pattern than the change in placebo, and goes in the opposite direction (placebo increases while $D2$ decreases).

These contrasts belong to the **Drug:Diet** interaction factor, and from the degrees of freedom in the Hasse diagram we know that two orthogonal contrasts partition its sum of squares into individual contributions. However, our two interaction contrasts are *not* orthogonal and therefore do not partition the interaction sum of squares.

Two alternative interaction contrasts that are orthogonal use weight vectors $\mathbf{w}_7 = (0, -1, 1, 0, 1, -1)$ and $\mathbf{w}_8 = (1, -1/2, 1/2, -1, -1/2, +1/2)$, respectively. The first contrasts the effects of the two active drugs under the two diets, while the second compares the placebo treatment to the average of the two active drugs.

The estimates for these two contrasts are shown in Table 6.7. They show that the difference in enzyme levels between the two diets is much larger for $D1$ than for $D2$, and much lower for placebo compared to the average of the two drugs. How far the latter contrast is biologically meaningful is of course another question, and our first set of non-orthogonal contrasts is likely more meaningful and easier to interpret.

Together with the three contrasts $\Psi(\mathbf{w}_1)$, $\Psi(\mathbf{w}_2)$, and $\Psi(\mathbf{w}_3)$ defined in Sect. 6.6.1, which individually contrast each drug between the two diets, these two orthogonal interaction contrasts form a set of five orthogonal contrasts that fully exhaust the information in the data. In principle, the three treatment F-tests of the two main drug and diet effects and the drug-by-diet interaction can be reconstituted from the F-tests of the interaction sums of squares.

6.7 Power Analysis and Sample Size

The power analysis and determination of sample sizes for omnibus F-tests in a multi-way ANOVA follow the same principles as for a one-way ANOVA and are based on the noncentral F-distribution. Similarly, the power analysis for linear contrasts is identical to the one-way ANOVA case. The ideas of portable power can also be applied, so we give only some additional remarks.

6.7.1 Main Effects

To determine the power of an omnibus F-test for a main effect of **A**, we need the significance level α, the effect size (such as f_A^2 or $\eta_{p,A}^2$), and the residual variance σ_e^2. The sample size n_A is now the number of samples per level of **A**, and is distributed evenly over the levels of all other factors crossed with **A**. For our generic two-way design with factors **A** and **B** with a and b levels, respectively, the number of samples per treatment combination is then $n = n_A/b$. The degrees of freedom for the denominator are $\mathrm{df}_{res} = ab(n-1)$ for the full model with interaction, and $\mathrm{df}_{res} = nab - a - b + 1$ for the additive model; we directly find them from the experiment diagram.

For our drug–diet example, we consider $n_B = 12$ mice per diet and thus $n = 12/3 = 4$ mice per drug–diet combination, a significance level of $\alpha = 5\%$ and use our previous estimate for the residual variance $\hat{\sigma}_e^2 = 1.5$. These parameters provide a power of $1 - \beta = 64\%$ for detecting a minimal effect size of $d_0 = 1$ or equivalently $f_0^2 = d_0^2/4 = 1/4$.

As a second example, we consider a 'medium' standardized main drug effect of $f_0 = 0.25$ and a desired power of 80% at a 5% significance level. This requires a sample size of $n_A = 53$ mice per drug level, about $n = 27$ mice per drug–diet treatment combination. The standardized difference between two drug averages is then $d_0 = \sqrt{f_0^2 \cdot 6} = 0.61$, corresponding to a raw difference in enzyme levels of $\delta_0 = d_0 \cdot \sigma_e = 0.75$ for the assumed residual variance.

Extending these ideas to the main effects of higher-order factorial designs is straightforward and only requires that we correctly calculate the residual degrees of freedom to take account of the other factors in the design. The resulting sample size is then again per level of the factor of interest, and has to be divided by the product of the number of levels of all other factors to get to the sample size per treatment. For example, with a three-way design with $a = b = 4$ and $c = 5$ factor levels, a resulting sample size of $n_A = 60$ for an **A** main effect means that we need $n = n_A/bc = 3$ samples per treatment combination.

6.7.2 Interactions

The denominator degrees of freedom for an F-test of an interaction factor are still the residual degrees of freedom, and we find the correct numerator degrees of freedom from the Hasse diagram. For our generic two-way example, these are $ab(n-1)$ and $(a-1)(b-1)$, respectively, and the resulting sample size corresponds to n directly. For higher-order factorial designs, we again have to divide the resulting sample size by the product of levels of the factors not involved in the interaction.

The standardized effect size for an interaction **A:B** is $f^2 = \sum_i \sum_j (\alpha\beta)_{ij}^2 / ab\sigma_e^2$, but defining a minimal effect size that corresponds to a scientifically meaningful interaction can be challenging. Three options are (i) to define the interaction parameters $(\alpha\beta)_{ij}$ directly, (ii) define the expected responses μ_{ij} for each combination of levels of **A** and **B** and work out the parameters from there, and (iii) use information about the main effects from previous single-factor studies and define a level of attenuation of these effects due to the second factor.

We look into the third option more closely and denote the two main effect sizes of previous single-factor studies by $f_{A,\,single}^2$ and $f_{B,\,single}^2$; note that the main effects will likely change in a two-way design. The same situation occurs if we have reasonable guesses for the two isolated main effect sizes. We consider how much attenuation or increase in the difference in response we expect for **A**, say, when looking at the low level of **B** compared to its high level. The effect size of the interaction is then about the same size as the single main effects if we expect a complete effect reversal, and we can use $f_{A,\,single}^2$ as a reasonable expected effect size for the interaction (Fig. 6.5D).

If we expect full attenuation of the effect of one factor, such as in Fig. 6.5C, then the interaction effect size is about one-half the main effect size. With lower attenuation, the interaction effect decreases accordingly. From these considerations, we find a crude guess for the required sample size for detecting an interaction effect: the total sample size is roughly the same as for the original main effect for a complete effect reversal. If we needed n samples for detecting a certain minimal **A** main effect with given power, then we need again n samples for each level of **A**, which means n/b samples per treatment. If we expect full attenuation, the total required sample size doubles, and for a 50% attenuation, we are already at four times the sample size that we would need to detect the single-factor main effect. The arguments become much more involved for interactions of factors with more than two levels.

We can alternatively forsake a power analysis for the interaction omnibus F-test, and instead specify (planned) interaction contrasts. The problem of power and sample size is then shifted to corresponding analyses for contrasts, whose minimal effect sizes are often easier to define.

6.7.3 Contrasts

The power analysis for contrasts is based on the same methods as for the one-way design discussed in Sect. 5.2.8. For a two-way design with a and b levels, we can find $ab - 1$ orthogonal contrasts, $a - 1$ for the first main effect, $b - 1$ for the second, and $(a - 1)(b - 1)$ to partition the interaction. A sensible strategy is to define the planned contrasts for the experiment, use power analysis to determine the required sample size for each contrast, and then use the largest resulting sample size for implementing the experiment.

If the contrasts are not orthogonal, we should consider adjustments for multiple testing. Since we do not have the resulting p-values yet, the single-step Bonferroni correction with significance levels α/q for a set of q contrasts is the only real choice for a set of general-purpose contrasts.

6.7.4   Portable Power

We note that the noncentrality parameter is again the product of effect size and *total* size of the experiment ($\lambda = n \cdot a \cdot b \cdot f^2$ for a two-factor ANOVA), and we can use the portable power formula $n = \phi^2/f^2$ to determine the number of experimental units *per level of the factor considered*.

For our drug–diet example with three drugs and two diets, the necessary sample size for detecting a medium drug main effect of at least $f_{\text{drug}} = 0.25$ with 80% power and significance level $\alpha = 5\%$ is $n_{\text{drug}} = \phi^2/0.25^2 = 48$ mice *per drug* (using $\phi^2 = 3$). This translates to 24 mice per treatment in good agreement with our previous exact calculations in Sect. 6.7.1.

A useful shortcut exists for the most frequent type of interaction contrast which uses two levels each of m treatment factors, with weights $w_i = \pm 1$ for two chosen levels of each factor, and zero otherwise. The weight vector of such contrast has 2^m non-zero entries, hence the estimator variance is $\sigma_e^2/n \cdot 2^m$. For detecting a minimal contrast size of Ψ_0 or a minimal distance δ_0 between any two cell means in the contrast, the required sample size is

$$n = \frac{\phi^2}{\Psi_0^2/\sigma_e^2} \cdot 2^{m-3} \quad \text{respectively} \quad n = \frac{\phi^2}{\delta_0^2/\sigma_e^2} \cdot \frac{1}{2^{m-3}} \ .$$

The sample size n is the number of samples for each combination of the treatment factors involved in the interaction.

6.8   Notes and Summary

Notes

The term *factorial design* was introduced by Fisher in the first edition of his book *Design of Experiments* (Fisher 1971) in 1935. The advantages of this design are universally recognized and re-iterated regularly for specific fields such as animal experiments (Shaw and Festing 2002).

Interpretation of interactions is a specific problem in factorial designs, and depends on the type of factors involved in an interaction (de Gonzalez and Cox 2007). Two interesting perspectives are found in Finney (1948) and Abelson and Prentice (1997). 'Spurious' interactions can arise in factorial designs with heterogeneous variance (Snee 1982).

Due to the potentially large number of treatment level combinations, methods for factorial designs without replication are required. An additive model allows pooling interactions and residual factors (Xampeny et al. 2018), and several methods for detecting additivity are compared in Rusch (2009). A general discussion of pooling in ANOVA is Hines (1996). The short note in Lachenbruch (1988) discusses sample sizes for testing interactions.

Robust methods such as Kruskal–Wallis do not readily extend to the multi-factor case, and naive application can lead to incorrect analysis (Mee and Lu 2010).

Analysis of unbalanced data in factorial designs is a long-standing problem first reviewed in detail in Yates (1934). Introductory treatments with simple examples are Urquhart and Weeks (1978), Shaw and Mitchell-Olds (1993), and Hector et al. (2010). A historic perspective (itself 30 years old now) is Herr (1986). An excellent treatment of the subject is Searle (1987). Nowadays, many of these problems are solved using linear mixed models instead of classical ANOVA; we discuss these in more detail in Sect. 7.2.2.

Using R

We can use the same set of functions for a multi-way ANOVA as for the one-way ANOVA. For contrast analysis, `emmeans()` provides a convenient shortcut if contrasts are defined for only one factor, and are to be evaluated independently within each level of the other factor. Our first set of contrasts can be alternatively calculated by defining contrasts based on three weights (one per drug), and then using `emmeans(m, ~drug|diet)` for evaluating these for both diets.

Due to the expanding possibilities for defining designs with multiple factors, there are no built-in functions for power analysis. The package `Superpower` provides the function `ANOVA_power()` for power analysis of single- and multi-factor ANOVAs, and interfaces directly with `emmeans` for power analysis of corresponding contrasts. The `design.crd()` function from package `agricolae` allows designing and randomizing CRDs.

Tukey's additivity test is implemented in the functions `tukey.1df()` from package `dae` and `tukey.test()` from package `additivityTests`.

The functions `cohens_f()` and `eta_squared()` from package `effectsize` extend to multi-way ANOVA results. Both provide exact confidence intervals and have an option `partial=` to compute the partial variant of the effect size.

Summary

Factorial treatment designs apply combinations of levels of several treatment factors to experimental units. This allows the detection and estimation of interactions, where the effect of one factor depends on the level of another factor. Even without such interactions, factorial designs are more efficient than multiple one-variable-at-a-time designs to study the main effects of several treatment factors. Simple designs can often be augmented with additional treatment factors: this allows studying more effects in the same experiment, and also broadens the experimental conditions under which each individual factor is studied, thereby increasing the generalizability of results.

Factorial designs are analyzed by multi-way ANOVA or similar methods. The importance of a factor can be gauged from its associated F-test, potentially allowing model reduction with adherence to the marginality principle. Linear contrasts are equivalent to the one-factor case and can be conveniently based on the estimated marginal means of the treatment groups. Care is required when interpreting results in the presence of interactions.

References

Abelson, R. P. and D. A. Prentice (1997). "Contrast tests of interaction hypothesis". In: Psychological Methods 2.4, pp. 315–328.

Davies, O. L. (1954). The Design and Analysis of Industrial Experiments. Oliver & Boyd, London.

de Gonzalez, A. and D. R. Cox (2007). "Interpretation of interaction: a review". In: The Annals of Applied Statistics 1.2, pp. 371–385.

Finney, D. J. (1948). "Main Effects and Interactions". In: Journal of the American Statistical Association 43.244, pp. 566–571.

Fisher, R. A. (1926). "The Arrangement of Field Experiments". In: Journal of the Ministry of Agriculture of Great Britain 33, pp. 503–513.

Fisher, R. A. (1971). The Design of Experiments. 8th. Hafner Publishing Company, New York.

Hector, A., S. von Felten, and B. Schmid (2010). "Analysis of variance with unbalanced data: An update for ecology & evolution". In: Journal of Animal Ecology 79.2, pp. 308–316.

Herr, D. G. (1986). "On the History of ANOVA in Unbalanced, Factorial Designs: The First 30 Years". In: The American Statistician 40.4, pp. 265–270.

Hines, W. G. S. (1996). "Pragmatics of Pooling in ANOVA Tables". In: The American Statistician 50.2, pp. 127–139.

Lachenbruch, P. A. (1988). "A note on sample size computation for testing interactions". In: Statistics in Medicine 7.4, pp. 467–469.

Lenth, R. V. (1989). "Quick and easy analysis of unreplicated factorials". In: Technometrics 31.4, pp. 469–473.

Littell, R. C. (2002). "Analysis of unbalanced mixed model data: A case study comparison of ANOVA versus REML/GLS". In: Journal of Agricultural, Biological, and Environmental Statistics 7.4, pp. 472–490.

Mee, R. W. and X. Lu (2010). "Don't use rank sum tests to analyze factorial designs". In: Quality Engineering 23.1, pp. 26–29.

Nelder, J. A. (1994). "The statistics of linear models: back to basics". In: Statistics and Computing 4.4, pp. 221–234.

Rusch, T. et al. (2009). "Tests of additivity in mixed and fixed effect two-way ANOVA models with single sub-class numbers". In: Statistical Papers 50.4, pp. 905–916.

Searle, S. R. (1987). Linear Models for Unbalanced Data. John Wiley & Sons, Inc.

Shaw, R. G. and T. Mitchell-Olds (1993). "Anova for unbalanced data: an overview". In: Ecology 74, pp. 1638–1645.

Shaw, R., M. F. W. Festing, et al. (2002). "Use of factorial designs to optimize animal experiments and reduce animal use". In: ILAR Journal 43.4, pp. 223–232.

Snee, R. D. (1982). "Nonadditivity in a two-way classification: Is it interaction or nonhomogeneous variance?" In: Journal of the American Statistical Association 77.379, pp. 515–519.

Tukey, J. W. (1949b). "One Degree of Freedom for Non-Additivity". In: Biometrics 5.3, pp. 232–242.

Urquhart, N. S. and D. L. Weeks (1978). "Linear Models in Messy Data: Some Problems and Alternatives". In: Biometrics 34.4, pp. 696–705.

Venables, W. N. (2000). Exegeses on linear models. Tech. rep.

Xampeny, R., P. Grima, and X. Tort-Martorell (2018). "Selecting significant effects in factorial designs: Lenth's method versus using negligible interactions". In: Communications in Statistics - Simulation and Computation 47.5, pp. 1343–1352.

Yates, F. (1934). "The Analysis of Multiple Classifications with Unequal Numbers in the Different Classes". In: Journal of the American Statistical Association 29.185, pp. 51–66.

Chapter 7
Improving Precision and Power: Blocked Designs

7.1 Introduction

In our discussion of the vendor examples in Sect. 3.3, we observed huge gains in power and precision when allocating each vendor to one of two samples per mouse, contrasting the resulting enzyme levels for each pair of samples individually, and then averaging the resulting differences.

This design is an example of *blocking*, where we organize the experimental units into groups (or *blocks*), such that units within the same group are more similar than those in different groups. For *k* treatments, a *block size* of *k* experimental units per group allows us to randomly allocate each treatment to one unit per block and we can estimate any treatment contrast within each block and then average these estimates over the blocks. Hence, this strategy removes any differences between blocks from the contrast estimates and tests, resulting in lower residual variance and increased precision and power without increases in sample size. As R. A. Fisher noted

> *Uniformity is only requisite between objects whose response is to be contrasted (that is, objects treated differently).* (Fisher 1971, p. 33)

To be effective, blocking requires that we find some property by which we can group our experimental units such that variances within each group are smaller than between groups. This property can be *intrinsic*, such as animal body weight, litter, or sex, or *non-specific* to the experimental units, such as day of measurement in a longer experiment, batch of some necessary chemical, or the machine used for incubating samples or recording measurements.

A simple yet powerful design is the *randomized complete block design (RCBD)*, where each block has as many units as there are treatments, and we randomly assign each treatment to one unit in each block. We can extend it to a *generalized randomized complete block design (GRCBD)* by using more than one replicate of each treatment per block. If the block size is smaller than the number of treatments, a *balanced incomplete block design (BIBD)* still allows treatment allocations balanced over blocks such that all pair contrasts are estimated with the same precision.

© Springer Nature Switzerland AG 2021
H.-M. Kaltenbach, *Statistical Design and Analysis of Biological Experiments*,
Statistics for Biology and Health, https://doi.org/10.1007/978-3-030-69641-2_7

Several blocking factors can be combined in a design by nesting—allowing estimation of each blocking factor's contribution to variance reduction—or crossing—allowing simultaneous removal of several independent sources of variation. Two classic designs with crossed blocks are *Latin squares* and *Youden squares*.

7.2   The Randomized Complete Block Design

7.2.1   Experiment and Data

We continue with our example of how three drug treatments in combination with two diets affect enzyme levels in mice. To keep things simple, we only consider the low-fat diet for the moment, so the treatment structure only contains **Drug** with three levels. Our aim is to improve the precision of contrast estimates and increase the power of the omnibus F-test. To this end, we arrange (or *block*) mice into groups of three and randomize the drugs separately within each group such that each drug occurs once per group. Ideally, the variance between animals in the same group is much smaller than between animals in different groups. A common choice for blocking mice is by litter, since sibling mice often show more similar responses as compared to mice from different litters (Perrin 2014). Litter sizes are typically in the range of 5–7 animals in mice (Watt 1934), which would easily allow us to select three mice from each litter for our experiment.

Our experiment is illustrated in Fig. 7.1A: we use $b = 8$ litters of $n = 3$ mice each, resulting in an experiment size of $N = bn = 24$ mice. In contrast to a completely randomized design, our randomization is *restricted* since we require that each treatment occurs exactly once per litter, and we randomize drugs independently for each litter.

The data are shown in Fig. 7.1B, where we connect the three observed enzyme levels in each block by a line, akin to an interaction plot. The vertical dispersion of the lines indicates that enzyme levels within each litter are systematically different from those in other litters. The lines are roughly parallel, which shows that all three drug treatments are affected equally by these systematic differences, there is no litter-by-drug interaction, and treatment contrasts are unaffected by systematic differences between litters.

7.2.2   Model and Hasse Diagram

Hasse Diagram

Deriving the experiment structure diagram in Fig. 7.2C poses no new challenges: the treatment structure contains the single treatment factor **Drug** with three levels

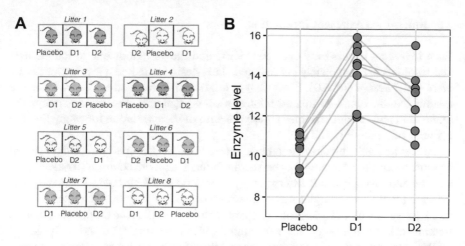

Fig. 7.1 Comparing enzyme levels for placebo, D1, and D2 under low-fat diet using randomized complete block design. **A** Data layout with eight litters of size three, treatments independently randomized in each block. **B** Observed enzyme levels for each drug; lines connect responses of mice from the same litter

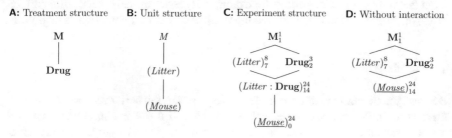

Fig. 7.2 Randomized complete block design for determining effect of two different drugs and placebo using eight mice per drug and a single measurement per mouse. Each drug occurs once per block and assignment to mice is randomized independently for each block

while the unit structure contains the blocking factor *(Litter)*, which groups the units from *(Mouse)*. We consider the specific eight litters in our experiment as a random sample, which makes both unit factors random. The factors **Drug** and *(Litter)* are crossed and their combinations are levels of the interaction factor *(Litter:**Drug**)*; since *(Litter)* is random, so is the interaction. The randomization allocates levels of **Drug** on *(Mouse)*, and the experiment structure is similar to a two-way factorial design, but *(Litter)* originates from the unit structure, is random, and is not randomly allocated to *(Mouse)*, but rather groups levels of *(Mouse)* by an intrinsic property of each mouse.

The Block-by-Treatment Interaction

Each treatment occurs only once per block, and the variations due to interactions and residuals are completely confounded. This design is called a *randomized complete block design (RCBD)*. If we want to analyze data from an RCBD, we need to assume that the block-by-treatment interaction is negligible. We can then merge the interaction and residual factors and use the sum of their variation for estimating the residual variance (Fig. 7.2D).

An appreciable block-by-treatment interaction means that the differences in enzyme levels between drugs depend on the litter. Treatment contrasts are then litter-specific and this systematic heterogeneity of treatment effects complicates the analysis and precludes a straightforward interpretation of the results. Such interaction is likely caused by other biologically relevant factors that influence the effect of (some) treatments but that have not been accounted for in our experiment and analysis.

We cannot test the interaction factor and therefore require a non-statistical argument to justify ignoring the interaction. Since we have full control over which property we use for blocking the experimental units, we can often employ subject-matter knowledge to exclude interactions between our chosen blocking factor and the treatment factor. In our particular case, for example, it seems unlikely that the litter affects drugs differently, which justifies treating the litter-by-drug interaction as negligible.

Linear Model

Because the blocking factor levels are random, so are the cell means μ_{ij}. Moreover, there is a single observation per cell and each cell mean is confounded with the residual for that cell; a two-factor cell means model is therefore not useful for an RCBD.

The parametric model (including block-by-treatment interaction, Fig. 7.2C) is

$$y_{ij} = \mu + \alpha_i + b_j + (\alpha b)_{ij} + e_{ij} ,$$

where μ is the grand mean, α_i the deviation of the average response for drug i compared to the grand mean, and b_j is the random effect of block j. We assume that the residuals and block effects are normally distributed as $e_{ij} \sim N(0, \sigma_e^2)$ and $b_j \sim N(0, \sigma_b^2)$, and are all mutually independent. The interaction of drug and block effects is a random effect with distribution $(\alpha b)_{ij} \sim N(0, \sigma_{\alpha b}^2)$.

For an RCBD, the interaction is completely confounded with the residuals; its parameters $(\alpha b)_{ij}$ cannot be estimated from the experiment without replicating treatments within each block. If interactions are negligible, then $(\alpha b)_{ij} = 0$ and we arrive at the additive model (Fig. 7.2D)

$$y_{ij} = \mu + \alpha_i + b_j + e_{ij} . \tag{7.1}$$

The average response to treatment level i is then $\mu_i = \mu + \alpha_i$; its unbiased estimator is

$$\hat{\mu}_i = \frac{1}{b} \sum_{j=1}^{b} y_{ij} \, ,$$

with variance $\text{Var}(\hat{\mu}_i) = (\sigma_b^2 + \sigma_e^2)/b$. We then base our contrast analysis on these estimated marginal means, effectively using a cell means model for the treatment factor after correcting for the block effects.

Analysis of Variance

The total sum of squares for the RCBD with an additive model decomposes into sums of squares for each factor involved, such that

$$\text{SS}_{\text{tot}} = \text{SS}_{\text{trt}} + \text{SS}_{\text{block}} + \text{SS}_{\text{res}} \, ,$$

and the degrees of freedom decompose accordingly.

The treatment factor is tested using the F-statistic

$$F = \frac{\text{SS}_{\text{trt}}/\text{df}_{\text{trt}}}{\text{SS}_{\text{res}}/\text{df}_{\text{res}}} \, ,$$

and mean squares are formed by dividing each sum of squares by its corresponding degrees of freedom.

For k treatment factor levels and b blocks, the test statistic has a $F_{k-1, (b-1)(k-1)}$-distribution under the omnibus null-hypothesis $H_0 : \mu_1 = \mu_2 = \cdots = \mu_k$ that all treatment group means are equal. Note that without blocking, the denominator sum of squares would be $\text{SS}_{\text{block}} + \text{SS}_{\text{res}}$ with corresponding loss of power.

We derive the model specification from the Hasse diagrams in Fig. 7.2: since fixed and random factors are exclusive to the treatment, respectively, unit structure, the corresponding terms are 1+drug and Error(litter/mouse); we reduce these terms to drug and Error(litter) and the model is fully specified as y~drug+Error(litter). The analysis of variance table then contains two *error strata*: one for the block effect and one for the within-block residuals.

We randomized the treatment on the mice within litters, and the within-block variance σ_e^2 therefore provides the mean squares for the F-test denominator; this agrees with the fact that *(Mouse)* is the closest random factor below **Drug** in the experiment structure diagram (Fig. 7.2D). The between-block error stratum contains no further factors or tests, since there are no factors randomized on *(Litter)* in this design. The ANOVA table is then

	Df	Sum Sq	Mean Sq	F value	Pr(>F)
Error stratum: Litter					
Residuals	7	37.08	5.3		
Error stratum: Within					
Drug	2	74.78	37.39	84.55	1.53e-08
Residuals	14	6.19	0.44		

The omnibus F-test for the treatment factor provides clear evidence that the drugs affect the enzyme levels differently and the differences in average enzyme levels between drugs are about 85 times larger than the residual variance.

The effect size $\eta_{drug}^2 = 63\%$ is large, but measures the variation explained by the drug effects compared to the overall variation, including the variation removed by blocking. It is more meaningful to compare the drug effect to the within-block residuals alone using the partial effect size $\eta_{p,\,drug}^2 = 92\%$, since the litter-to-litter variation is removed by blocking and has no bearing on the precision of the treatment effect estimates. The large effect sizes confirm that the overwhelming fraction of the variation of enzyme levels in each litter is due to the drugs acting differently.

In contrast to a two-way ANOVA with factorial treatment structure, we cannot simplify the analysis to a one-way ANOVA with a single treatment factor with $b \cdot k$ levels. This is because the blocking factor is random, and the resulting one-way factor would also be a random factor. The omnibus F-test for this factor is difficult to interpret, and a contrast analysis would be a futile exercise, since we would compare randomly sampled factor levels among each other.

Linear Mixed Model

In contrast to our previous designs, the analysis of an RCBD requires two variance components in our model: the residual within-block variance σ_e^2 and the between-block variance σ_b^2. The classical analysis of variance handles these by using one error stratum per variance component, but this makes subsequent analyses more tedious. For example, estimates of the two variance components are of direct interest, but while the estimate $\hat{\sigma}_e^2 = MS_{res}$ is directly available from the ANOVA result, an estimate for the between-block variance is

$$\hat{\sigma}_b^2 = \frac{MS_{block} - MS_{res}}{n} \, ,$$

and requires manual calculation. For our example, we find $\hat{\sigma}_e^2 = 0.44$ and $\hat{\sigma}_b^2 = 1.62$ based on the ANOVA mean squares.

An attractive alternative is the *linear mixed model*, which explicitly considers the different random factors for estimating variance components and parameters of the linear model in Eq. (7.1). Linear mixed models offer a very general and powerful extension to linear regression and analysis of variance, but their general theory and estimation are beyond the scope of this book. For our purposes, we only need a small fraction of their possibilities and we use the `lmer()` function from package `lme4` for all our calculations. In specifying a linear mixed model, we use terms of the form `(1|X)` to introduce a random offset for each level of the factor `X`; this construct replaces the `Error()`-term from `aov()`. The fixed effect part of the model specification remains unaltered. For our example, the model specification is then `y~drug+(1|litter)`, which asks for a fixed effect (α_i) for each level of **Drug**, and allows a random offset (b_j) for each litter.

The `Fixed effects` section of the result gives the parameter estimates for α_i, but they are of secondary interest and depend on the coding (cf. Sect. 4.7).

	Estimate	Std. Error	df	t value	Pr(>|t|)
(Intercept)	10.01	0.51	9.4	19.73	5.73e-09
drugD1	4.23	0.33	14.0	12.71	4.46e-09
drugD2	2.91	0.33	14.0	8.74	4.83e-07

Note that the degrees of freedom for estimating the intercept (which here corresponds to the average enzyme level in the placebo group) is no longer an integer, because block effects and residuals are used with different weights in its estimation. The `Random effects` section of the result provides the variance and standard deviation for each variance component:

Group	Name	Variance	Std. Dev
Litter	(Intercept)	1.62	1.27
Residual		0.44	0.67

These correspond to the two variance components $\hat{\sigma}_b^2 = 1.62$ (`Litter`) and $\hat{\sigma}_e^2 = 0.44$ (`Residual`), in agreement with the ANOVA results.

We calculate our familiar ANOVA table from an estimated linear mixed model m using `anova(m)`, which yields

	Sum Sq	Mean Sq	NumDF	DenDF	F value	Pr(>F)
Drug	74.78	37.39	2	14	84.55	1.53e-08

Linear mixed models and 'traditional' analysis of variance use the same linear model to analyze data from a given design. Their main difference is the way they handle models with multiple variance components. Analysis of variance relies on the crude concept of error strata, which makes direct estimation of variance more cumbersome and leads to loss in efficiency if information on effects is distributed between error strata (such as in incomplete block designs, discussed in Sect. 7.3). Linear mixed models use different techniques for estimation of the model's parameters that make use of all available information. Variance estimates are directly available, and linear mixed models do not suffer from problems with unbalanced group sizes. Colloquially, we can say that the analysis of variance approach provides a convenient framework to phrase design questions, while the linear mixed model provides a more general tool for the subsequent analysis.

7.2.3   Contrasts

In a blocked design with random block factor, the only useful contrasts are between levels of (fixed) treatment factors, since a contrast involving levels of the blocking factor would define a comparison of specific instances of random factor levels. To estimate the difference between the average enzyme level of the first and second litter, for example, only has a useful interpretation if these two arbitrary litters are again used in a replication of the experiment. This is also reflected in the population marginal means, of which there are only three—$\mu_1, \ldots, \mu_3$, one per treatment group—in our example, rather than one per block-treatment combination. We then define and estimate treatment contrasts exactly as before, and briefly exemplify the procedure for our experiment with three contrasts: the comparison of $D1$, respectively, $D2$ to placebo, and the comparison of the average of $D1$ and $D2$ to placebo. The contrasts are

$$\Psi(\mathbf{w}_1) = \mu_2 - \mu_1, \quad \Psi(\mathbf{w}_2) = \mu_3 - \mu_1, \quad \text{and} \quad \Psi(\mathbf{w}_3) = \frac{\mu_2 + \mu_3}{2} - \mu_1 .$$

The following code defines and estimates the contrasts based on the linear mixed model:

```
# Estimate linear mixed model
lmm.rcbd = lmer(y~drug+(1|litter), data=drugs_litters)
# Estimate group means
em.lmm = emmeans(lmm.rcbd, ~drug)
# Manually define contrasts
psi = list("D1 - Placebo"=c(-1,+1,0),
           "D2 - Placebo"=c(-1,0,+1),
           "(D1+D2)/2 - Placebo"=c(-1,+1/2,+1/2))
# Estimate contrasts
ct.lmm= contrast(em.lmm, psi)
# Get 95%-CI
ci.lmm = confint(ct.lmm)
```

The contrast estimates for our experiment are shown in Table 7.1 for an analysis based on `aov()`, the above analysis using `lmer()`, and an incorrect analysis of variance that ignores the blocking factor and treats the data as coming from a completely randomized design. Ignoring the blocking factor results in a residual variance of $\sigma_b^2 + \sigma_e^2$ instead of σ_e^2 for the correct analysis. The resulting decrease in precision is clearly visible.

Table 7.1 Contrast estimates and confidence intervals based on three linear models

Contrast	Estimate	se	df	LCL	UCL
ANOVA: aov(y~drug+Error(litter))					
D1 – Placebo	·4.23	0.33	14	3.51	4.94
D2 – Placebo	2.91	0.33	14	2.19	3.62
(D1+D2)/2 – Placebo	3.57	0.29	14	2.95	4.18
Mixed model: lmer(y~drug+(1|litter))					
D1 – Placebo	4.23	0.33	14	3.51	4.94
D2 – Placebo	2.91	0.33	14	2.19	3.62
(D1+D2)/2 – Placebo	3.57	0.29	14	2.95	4.18
INCORRECT: aov(y~drug)					
D1 – Placebo	4.23	0.72	21	2.73	5.72
D2 – Placebo	2.91	0.72	21	1.41	4.40
(D1+D2)/2 – Placebo	3.57	0.62	21	2.27	4.86

7.2.4 Evaluating and Choosing a Blocking Factor

Evaluating a Blocking Factor

Once the data are recorded, we are interested in quantifying how 'good' the blocking performed in the experiment. This information would allow us to better predict the expected residual variance for a power analysis of our next experiment and to determine if we should continue using the blocking factor.

One way of evaluating the blocking is by treating the blocking factor as fixed, and specify the ANOVA model as y~drug+litter. The ANOVA table then contains a row for the blocking factor with an associated F-test (Samuels et al. 1991). This test only tells us if the between-block variance is significantly different from zero, but does not quantify the advantage of blocking.

A more meaningful alternative is the calculation of appropriate effect sizes, and we can determine the percentage of variation removed from the analysis by blocking using the effect size η^2_{block}, which evaluates to 31% for our example. In addition, the partial effect $\eta^2_{p,\,block} = 86\%$ shows that the vast majority of non-treatment sum of squares is due to the litter-to-litter variation and blocking by litter was a very successful strategy.

Alternatively, the *intraclass correlation coefficient* ICC $= \sigma^2_b/(\sigma^2_b + \sigma^2_e)$ uses the proportion of variance and we find an ICC of 79% for our example, confirming that the blocking works well. The ICC is directly related to the relative efficiency of the RCBD compared to a non-blocked CRD, since

$$\text{RE(CRD, RCBD)} = \frac{\sigma^2_b + \sigma^2_e}{\sigma^2_e} = \frac{1}{1 - \text{ICC}}, \tag{7.2}$$

and $\sigma_b^2 + \sigma_e^2$ is the residual variance of a corresponding CRD. From our estimates of the variance components, we find a relative efficiency of 4.66; to achieve the same precision as our blocked design with 8 mice per treatment group would require about 37 mice per treatment group for a completely randomized design.

We can alternatively find the relative efficiency as

$$\text{RE(CRD, RCBD)} = \frac{\widehat{MS}_{\text{res, CRD}}}{MS_{\text{res, RCBD}}},$$

where we estimate the residual mean squares of the CRD from the ANOVA table of the RCBD as a weighted average of the block and the residual sums of squares, weighted by the corresponding degrees of freedom:

$$\widehat{MS}_{\text{res, CRD}} = \frac{df_{\text{block}} \cdot MS_{\text{block}} + (df_{\text{trt}} + df_{\text{res}}) \cdot MS_{\text{res}}}{df_{\text{block}} + df_{\text{trt}} + df_{\text{res}}}.$$

For our example, $\widehat{MS}_{\text{res, CRD}} = 1.92$, resulting in a relative efficiency estimate of RE(CRD,RCBD) = 4.34. It differs slightly from our previous estimate because it uses an approximation based on the ANOVA results rather than estimates of the variance components based on the linear mixed model.

Choosing a Blocking Factor

The main purpose of blocking is to decrease the residual variance for improving the power of tests and precision of estimates by grouping the experimental units *before* the allocation of treatments, such that units within the same block are more similar than units from different blocks. The blocking factor must therefore describe a property of experimental units that affects all treatments equally to avoid a block-by-treatment interaction, and that is independent of the treatment outcome, which precludes, e.g., grouping 'responders' and 'non-responders' based on the measured response.

Some examples of grouping units using *intrinsic properties* are (i) litters of animals; (ii) parts of a single animal or plant, such as left/right kidney or leaves; (iii) initial weight; (iv) age or co-morbidities; (v) biomarker values, such as blood pressure, state of tumor, or vaccination status. Properties *non-specific to the experimental units* include (i) batches of chemicals used for the experimental unit; (ii) device used for measurements; or (iii) date in multi-day experiments. These are often necessary to account for systematic differences from the logistics of the experiment (such as batch effects); they also increase the generalizability of inferences due to the broader experimental conditions.

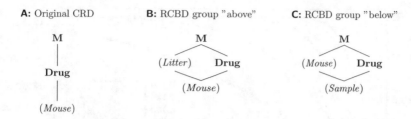

Fig. 7.3 Creating a randomized complete block design from a completely randomized design. **A** A CRD randomizes drugs on mice. **B** Introducing a blocking factor 'above' groups the experimental units. **C** Subdividing each experimental units and randomizing the treatments on the lower level creates a new experimental unit factor 'below'

The golden rule for designing experiments with blocks is

Block what you can, randomize what you can not.

As always, practical considerations should be taken into account when deciding upon blocking and randomization. There is no point in designing an overly complicated blocked experiment that becomes too difficult to implement correctly. On the other hand, there is no harm if we use a blocking factor that turns out to have no or minimal effect on the residual variance.

We can think about creating a blocked design by starting from a completely randomized design and 'splitting' the experimental unit factor into a blocking and a nested (potentially new) unit factor. Two examples are shown in Fig. 7.3, starting from the CRD (Fig. 7.3A) randomly allocating drug treatments on mice. In the first RCBD (Fig. 7.3B), we create a blocking factor 'above' the original experimental unit factor and group mice by their litters. This restricts the randomization of **Drug** to mice within litters. In the second RCBD (Fig. 7.3C), we subdivide the experimental unit into smaller units by taking multiple samples per mouse. This re-purposes the original experimental unit factor as the blocking factor and introduces a new factor 'below', but requires that we now randomize **Drug** on *(Sample)* to obtain an RCBD and not pseudo-replication.

7.2.5 Power Analysis and Sample Size

Power analysis for a randomized complete block design with no block-by-treatment interaction is very similar to a completely randomized design, the only difference being that we use the within-block variance σ_e^2 as our residual variance. For a main effect with parameters α_i, the noncentrality parameter is thus $\lambda_R = n \sum_i \alpha_i^2 / \sigma_e^2 = n \cdot k \cdot f_R^2$, while it is $\lambda_C = n \sum_i \alpha_i^2 / (\sigma_b^2 + \sigma_e^2) = n \cdot k \cdot f_C^2$ for the same treatment structure in a completely randomized design. The two effect sizes are different for the same raw effect size, and f_C^2 is smaller than f_R^2 for the same values of n, k, and α_i; this directly reflects the increased precision in an RCBD.

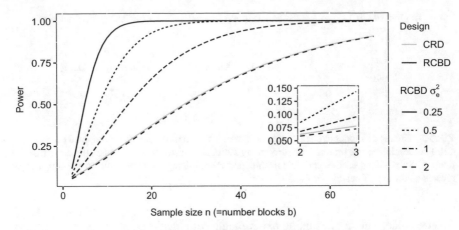

Fig. 7.4 Power for different sample sizes n for a completely randomized design with residual variance two and four randomized complete block designs with same overall variance, but four different within-block residual variances. Inset: same curves for small sample sizes show lower power of RCBD than CRD for identical residual variance due to lower residual degrees of freedom in the RCBD

The denominator degrees of freedom are $(k-1)(b-1) = kb - k - (b-1)$ for an RCBD, and we loose $b-1$ degrees of freedom compared to the $k(b-1) = kb - k$ degrees of freedom for a CRD. The corresponding loss in power is very small and more than compensated by the simultaneous increase in power due to the smaller residual variance unless the blocking is very ineffective and the number of blocks large.

As an example, we consider a single treatment factor with $k = 3$ levels, a total sample size of $N = b \cdot k = 30$, and an effect size of $f_{C,0}^2 = 0.0625$ for the treatment main effect at a significance level of $\alpha = 5\%$. The residual variance is $\sigma^2 = 2$. Under a completely randomized design, we have a power of 20% for detecting the effect. This power increases dramatically to 61% when we organize the experimental units into $b = 10$ blocks with between-block variance $\sigma_b^2 = 1.5$ and within-block variance $\sigma_e^2 = 0.5$. The much smaller residual variance increases the standardized effect size to $f_{R,0}^2 = (\sigma_b^2 + \sigma_e^2)/\sigma_e^2 \cdot f_{C,0}^2 = 0.25$; the increase is exactly the relative efficiency (Eq. (7.2)).

The relation between power and within-block variance is shown in Fig. 7.4 for five scenarios with significance level $\alpha = 5\%$ and an omnibus-F test for a main effect with $k = 3$ treatment levels. The solid grey curve shows the power for a CRD with residual variance $\sigma_e^2 + \sigma_b^2 = 2$ as a function of sample size per group and a minimal detectable effect size of $f_C^2 = 0.0625$. The black curves show the power for an RCBD with four different within-block variances from $\sigma_e^2 = 0.25$ to $\sigma_e^2 = 2$. For the same sample size, the RCBD then has larger effect size f_R^2, which translates into higher power. The inset figure shows that an RCBD has negligibly *less* power than a CRD for the same residual variance of $\sigma_e^2 = 2$, due to the loss of $b-1$ degrees of freedom.

A: Treatment structure   **B:** Unit structure   **C:** Experiment structure

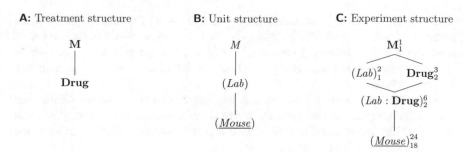

Fig. 7.5 A (generalized) randomized complete block design (GRCBD) with four mice per drug and laboratory. Completely randomized design for estimating drug effects of three drugs replicated in two laboratories

7.2.6 Replication Within Blocks

Within-block replication by using each treatment multiple times in each block is usually not a profitable strategy, and we should increase replication in an RCBD by introducing more blocks. A rare exception occurs when the blocking factor is not very efficient or measurement error is large, and the residual variance σ_e^2 is large compared to the between-block variance σ_b^2. Observing each treatment multiple times per block might then increase the precision of within-block contrast estimates.

In some scenarios, however, it is necessary to use more than one replicate of each treatment per block. This is typically the case if our blocking factor is non-specific and introduced to capture grouping due to the logistics of the experiment. For example, we might conduct our drug comparisons in two different laboratories. We then replicate a full CRD in the two laboratories, leading to a design with *(Lab)* as a blocking factor, but multiple observations of each treatment in each laboratory. This design is helpful to account for laboratory-specific differences.

Another purpose for using a multi-laboratory experiment is to broaden the inference for our experiment. If we find stable estimates of contrasts between laboratories despite the expected systematic differences, then we demonstrated experimentally that our scientific findings are independent of 'typical' alterations in the protocol or chemicals used, for example.

The resulting design is called a *generalized randomized complete block design (GRCBD)* and its diagrams are shown in Fig. 7.5 for eight mice per drug and two laboratories

Each mouse is used in exactly one laboratory and *(Lab)* thus groups *(Mouse)*. We use the same drugs in both laboratories, making *(Lab)* and **Drug** crossed factors. The replication of treatments in each block allows estimation of the lab-by-drug interaction, which provides information about the stability of the effects and contrasts. The linear model with interaction is

$$y_{ijk} = \mu + \alpha_i + b_j + (\alpha b)_{ij} + e_{ijk} ,$$

where α_i are the (fixed) treatment effect parameters, b_j the random block effect parameters with $b_j \sim N(0, \sigma_b^2)$, $(\alpha b)_{ij} \sim N(0, \sigma_{\alpha b}^2)$ describes the interaction of treatment i and laboratory j, and $e_{ijk} \sim N(0, \sigma_e^2)$ are the residuals. The specification for a linear mixed model is y ~ drug + (1|lab) + (1|lab:drug) for our example.

If the lab-by-treatment interaction is negligible with $\sigma_{\alpha b}^2 \approx 0$, we remove the interaction term from the analysis and arrive at the additive model. Then, *(Lab)* acts as a blocking factor just like *(Litter)* in the previous example and eliminates the lab-to-lab variation from treatment contrasts.

If the interaction is appreciable and cannot be neglected, then the contrasts between drugs are different between the two labs, and we cannot determine generalizable drug effects. Now, *(Lab:***Drug***)* is the first random factor below **Drug** and provides the factor for comparing levels of **Drug**, and *(Mouse)* is sub-sampled from *(Lab:***Drug***)*. Consequently, the variation against which drug effects are tested is larger and replication is lower, leading to decrease in precision and power for treatment contrasts.

7.2.7 Fixed Blocking Factors

We can typically consider the levels of a blocking factor as randomly drawn from a set of potential levels. The blocking factor is then random, and we are not interested in contrasts involving its levels, for example, but rather use the blocking factor to increase precision and power by removing parts of the variation from treatment contrasts.

A typical example of a non-random classification factor is the sex of an animal. The Hasse diagrams in Fig. 7.6 show an experiment design to study the effects of our three drugs on both female and male mice. Each treatment group has eight mice, with half of them female, the other half male. The experiment design looks similar to a factorial design of Chap. 6, but the interpretation of its analysis is rather different. Most importantly, while the factor *Sex* is fixed with only two possible levels, its levels are not randomly assigned to mice. This is reflected in the fact that *Sex* groups mice by an intrinsic property and hence belongs to the unit structure. In contrast, levels of **Drug** are randomly assigned to mice, and **Drug** therefore belongs to the treatment structure of the experiment.

In contrast to a randomized complete block design, we cannot increase the number of blocks to increase the replication. With two levels of *Sex* fixed, we instead need to increase the experiment size by using multiple mice of each sex for each drug. Since *Sex* and **Drug** are fixed, so is their interaction, and all fixed factors are tested against the same error term from *(Mouse)*.

The specification for the analysis of variance is y~drug*sex, and it looks exactly like the specification of a two-way ANOVA. However, since *Sex* is a unit factor that is not randomized, the roles of *Sex* and **Drug** are not symmetric and the interpretation is rather different from our previous two-way ANOVA with treatment factors **Drug** and **Diet**. In the factorial design, the presence of a **Drug:Diet** interaction means that

A: Treatment structure   **B:** Unit structure   **C:** Experiment structure

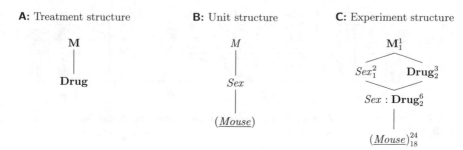

Fig. 7.6 A blocked design using sex of mouse as a fixed classification factor, with four female and four male mice in of three each treatment groups

we can modify the effect of a drug by using a different diet. In the fixed block design, a *Sex*:**Drug** interaction means that the effect of drugs is not stable over the sexes and some drugs affect female and male mice differently. We can use a drug to alter the enzyme level in a mouse, but we cannot alter the sex of that mouse to modulate the drug effect.

7.2.8 Factorial Treatment Structure

It is straightforward to extend an RCBD from a single treatment factor to factorial treatment structures by crossing the entire treatment structure with the blocking factor. Each block then contains one full replicate of the factorial design, and the required block size rapidly increases with the number of factors and factor levels. In practice, only smaller factorials can be blocked by this method since heterogeneity between experimental units often increases with block size, diminishing the advantages of blocking. We discuss more sophisticated designs for blocking factorials that overcome this problem by using only a fraction of all treatment combinations per block in Chap. 9.

We illustrate the extension using our drug-diet example from Chap. 6 where three drugs were combined with two diets. The six treatment combinations can still be accommodated when blocking by litter, and using four litters of size six results in the experimental layout shown in Fig. 7.7 with 24 mice in total. Already including a medium-fat diet would require atypical litter sizes of nine mice, however.

Model and Hasse Diagram

For constructing the Hasse diagrams in Fig. 7.8, we note that the blocking factor is crossed with the full treatment structure—both treatment factors and the treatment interaction factor. We assume that our blocking factor does not differentially

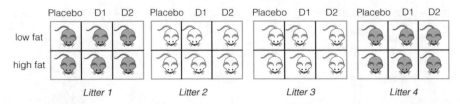

Fig. 7.7 Experiment layout for three drugs under two diets, four litter blocks. Two-way crossed treatment structure within each block

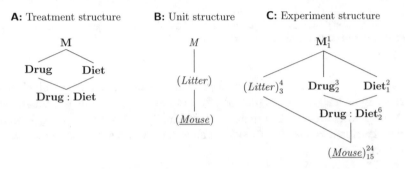

Fig. 7.8 Randomized complete block design for determining effect of two different drugs and placebo combined with two different diets using four mice per drug and diet, and a single response measurement per mouse. All treatment combinations occur once per block and are randomized independently

change the diet or drug effects and consequently ignore all three block-by-treatment interactions (*Litter*:**Diet**), (*Litter*:**Drug**), and (*Litter*:**Diet**:**Drug**).

We can easily derive a corresponding linear model from the Hasse diagrams. It is identical to our 'usual' model for an RCBD, except that its mean structure additionally contains the parameters for the second treatment factor and the interaction:

$$ y_{ijk} = \mu + \alpha_i + \beta_j + (\alpha\beta)_{ij} + b_k + e_{ijk} \, , $$

where α_i and β_j are the main effects of drug and diet, respectively, $(\alpha\beta)_{ij}$ the interaction effects between them, $b_k \sim N(0, \sigma_b^2)$ are the random block effects, and $e_{ijk} \sim N(0, \sigma_e^2)$ are the residuals. Again we assume that all random effects are mutually independent.

The enzyme levels y_{ijk} are shown in Fig. 7.9 as dark grey points in an interaction plot separately for each block. We can clearly see the block effects, which systematically shift enzyme levels for all conditions. After estimating the linear model, the shift b_k can be estimated for each block, and the light grey points show the resulting 'normalized' data $y_{ijk} - \hat{b}_k$ after accounting for the block effects.

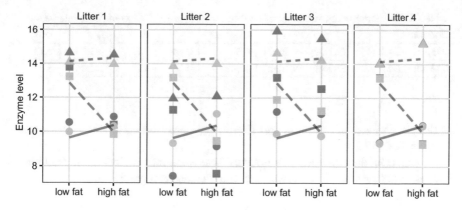

Fig. 7.9 Enzyme levels for placebo (point), drug D1 (triangle), and drug D2 (square). Data are shown separately for each of four litters (blocks). Lines connect mean values over litters of same drug under low and high-fat diets. Dark grey points are raw data, light grey points are enzyme levels adjusted for block effect

ANOVA Table

We analyze the resulting data using either analysis of variance or a linear mixed model. The two model specifications derive directly from the Hasse diagrams and consist of two main effects and interaction for the treatment factors and an error stratum or a random intercept for the litter unit factor. The specifications are y~drug*diet+Error(litter), respectively, y~drug*diet+(1| litter).

The resulting ANOVA table based on the linear mixed model is

	Sum Sq	Mean Sq	NumDF	DenDF	F value	Pr(>F)
Drug	74.1	37.05	2	15	91.54	3.93e-09
Diet	2.59	2.59	1	15	6.39	2.32e-02
Drug:Diet	15.41	7.71	2	15	19.04	7.65e-05

Blocking by litter reduces the residual variance from 1.19 for the non-blocked design in Chap. 6 to 0.64, the relative efficiency is RE(CRD,RCBD) = 1.87. Due to the increase in power, the **Diet** main effect is now significant.

Contrasts

The definition and analysis of linear contrasts work exactly as for the two-way ANOVA in Sect. 6.6, and contrasts are defined on the six treatment group means. For direct comparison with our previous results, we estimate the two interaction contrasts of Table 6.6 in the blocked design. They compare the difference in enzyme levels for D1 (resp. D2) under low and high-fat diet to the corresponding difference

Table 7.2 Interaction contrast estimates and Bonferroni-adjusted 95%-confidence intervals

Contrast	Estimate	se	df	LCL	UCL
(D1 low – D1 high) – (Pl. low – Pl. high)	0.54	0.64	15	−1.04	2.13
(D2 low – D2 high) – (Pl. low – Pl. high)	3.64	0.64	15	2.06	5.22

in the placebo group; estimates and Bonferroni-corrected confidence intervals are shown in Table 7.2.

As expected, the estimates are similar to those we found previously, but confidence intervals are substantially narrower due to the increase in precision.

7.3 Incomplete Block Designs

7.3.1 Introduction

The randomized complete block design can be too restrictive if the number of treatment levels is large or the available block sizes small. Fractional factorial designs offer a solution for factorial treatment structures by confounding some treatment effects with blocks (Chap. 9). For a single treatment factor, we can use *incomplete block designs (IBD)*, where we deliberately relax the complete balance of the previous designs and use only a subset of the treatments in each block.

The most important example is the *balanced incomplete block design (BIBD)*, where each pair of treatments is allocated to the same number of blocks, and so are therefore all individual treatments. This specific type of balance ensures that pairwise contrasts are all estimated with the same precision, but precision decreases if more than two treatment groups are compared.

We illustrate this design with our drug example, where three drug treatments are allocated to 24 mice on a low-fat diet. If the sample preparation is so elaborate that only two mice can be measured at the same time, then we have to run this experiment in twelve batches, with each batch containing one of three treatment pairs: (placebo, $D1$), (placebo, $D2$), or ($D1$, $D2$). A possible balanced incomplete block design is shown in Fig. 7.10, where each treatment pair is used in four blocks, and the treatments are randomized independently to mice within each block. The resulting data are shown in Table 7.3.

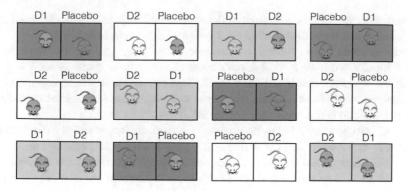

Fig. 7.10 Experiment layout for a balanced incomplete block design with three drugs in twelve batches of size two

Table 7.3 Data for a BIBD with 12 blocks of size 2 and 3 treatment levels

Batch	Drug	y	Batch	Drug	y	Batch	Drug	y
Block 1	Placebo	8.79	Block 5	D2	14.05	Block 9	D1	13.87
Block 1	D1	12.89	Block 5	Placebo	11.93	Block 9	D2	13.92
Block 2	D2	13.49	Block 6	D1	12.98	Block 10	Placebo	9.03
Block 2	Placebo	10.58	Block 6	D2	11.47	Block 10	D1	13.67
Block 3	D1	12.76	Block 7	Placebo	8.06	Block 11	D2	13.42
Block 3	D2	11.64	Block 7	D1	12.80	Block 11	Placebo	10.62
Block 4	Placebo	10.14	Block 8	D2	11.22	Block 12	D1	15.63
Block 4	D1	14.65	Block 8	Placebo	9.10	Block 12	D2	12.74

7.3.2 Defining a Balanced Incomplete Block Design

The key requirement of a BIBD is that *all pairs of treatments* occur the same number of times in a block. This requirement restricts the possible combinations of number of treatments, block size, and number of blocks. With three pairs of treatments as in our example, the number of blocks in a BIBD is restricted to be a multiple of three. The relations between the parameters of a BIBD are known as the two *defining equations*

$$rk = bs \quad \text{and} \quad \lambda \cdot (k - 1) = r \cdot (s - 1) . \tag{7.3}$$

They relate the number of treatments k, number of blocks b, block size s to the number of times r that each treatment occurs and the number of times λ that each pair of treatments occurs in the same block. If these equations hold, then the corresponding design is a BIBD.

The equations are derived as follows: with b blocks of s units each, the product bs is the total number of experimental units in the experiment. This number has to equal rk, since each of the k treatments occurs in r blocks. Moreover, each particular treatment occurs λ times with each of the remaining $k-1$ treatments. It also occurs in r blocks, and is combined with $s-1$ other treatments in each block.

For our example, we have $k=3$ treatment levels and each treatment occurs in $r=8$ blocks, we have $b=12$ blocks each of size $s=2$, resulting in $\lambda=4$ co-occurrences of each pair in the same block. This satisfies the defining equations and our design is a BIBD. In contrast, for only $b=5$ blocks, we are unable to satisfy the defining equations and no corresponding BIBD exists.

In order to generate a balanced incomplete block design, we need to find parameters s, r, and λ such that the defining Eq. (7.3) holds. A particularly simple way is by generating an *unreduced* or *combinatorial design* and use as many blocks as there are ways to select s out of k treatment levels. The number of blocks b is then $b = \binom{k}{s} = k!/(k-s)!s!$ and increases rapidly with k.

We can often improve on the unreduced design, and generate a valid BIBD with substantially fewer blocks. Experimental design books sometimes contain tables of BIBDs for a different number of treatment and block sizes, see for example Cochran and Cox (1957). As a further example, we consider constructing a BIBD for $k=6$ treatments in blocks of size $s=3$. The unreduced combinatorial design requires $b = \binom{6}{3} = 20$ blocks (experiment size $N=60$). Not all of these are needed to fulfill the defining equations, however, and a smaller BIBD exists with only $b=10$ blocks (and $N=30$).

7.3.3   Analysis

The balanced incomplete block design is not fully balanced: since only a fraction of the available treatment levels occurs in each block, some combinations of block and treatment levels are not observed, resulting in some-cells-empty data. The treatment and block main effects are no longer independent, but the requirements encoded in the two defining equations ensure that unbiased estimates exist for all treatment effects, and that standard ANOVA techniques are available for the analysis. However, the treatment and block factor sums of squares are not independent, and while their sum remains the same, their individual values now depend on the order of the two factors in the model specification. Moreover, part of the information about treatment effects is captured by the differences in blocks; in particular, there are two omnibus F-tests for the treatment factor, the first based on the *inter-block information* and located in the ANOVA table in the block-factor error stratum, and the second based—as for an RCBD—on the *intra-block information* and located in the residual error stratum. A linear mixed model automatically combines these information, and analysis of variance and mixed model results no longer concur exactly for a BIBD.

Linear Model

If we assume negligible block-by-treatment interaction, the linear model for a balanced incomplete block design is the additive model

$$y_{ij} = \mu + \alpha_i + b_j + e_{ij} \,,$$

where μ is the grand mean, α_i the expected difference between the group mean for treatment i and the grand mean, $b_j \sim N(0, \sigma_b^2)$ are the random block effects, and $e_{ij} \sim N(0, \sigma_e^2)$ are the residuals. All random variables are mutually independent.

Not all block-treatment combinations occur in a BIBD, and some of the y_{ij} therefore do not exist. As a consequence, naive parameter estimators are biased, and more complex estimators are required for the parameters and treatment group means. We forego a more detailed discussion of these problems and rely on statistical software like R to provide appropriate estimates for BIBDs.

Intra- and Inter-block Information

In an RCBD, we can estimate any treatment contrast and all effects independently within each block, and then average over blocks. We can use the same *intra-block analysis* for a BIBD by estimating contrasts and effects based on those blocks that contain sufficient information and averaging over these blocks. The resulting estimates are free of block effects.

In addition, the block totals—calculated by adding up all response values in a block—also contain information about contrasts and effects if the block factor is random. This information can be extracted by an *inter-block analysis*. We again refrain from discussing the technical details, but provide some intuition for the recovery of inter-block information.

We consider our example with three drugs in blocks of pairs and are interested in estimating the average enzyme level under the placebo treatment. Using the intra-block information, we would adjust each response value by the block effect, and average over the resulting adjusted placebo responses. For the inter-block analysis, first note that the true treatment group means are $\mu + \alpha_1$, $\mu + \alpha_2$, and $\mu + \alpha_3$ for placebo, $D1$, and $D2$, respectively. The block totals for the three types of block are then (placebo, $D1$)= $T_{P,D1} = \mu + \alpha_1 + \mu + \alpha_2$, (placebo, $D2$)= $T_{P,D2} = \mu + \alpha_1 + \mu + \alpha_3$, and ($D1$, $D2$)= $T_{D1,D2} = \mu + \alpha_2 + \mu + \alpha_3$, respectively, and each type of block occurs the same number of times.

The treatment group mean for placebo can then be calculated from the block totals as $(T_{P,D1} + T_{P,D2} - T_{D1,D2})/2$, since

$$\frac{T_{P,D1} + T_{P,D2} - T_{D1,D2}}{2} = \frac{2\mu + \alpha_1 + \alpha_2}{2} + \frac{2\mu + \alpha_1 + \alpha_3}{2} - \frac{2\mu + \alpha_2 + \alpha_3}{2}$$
$$= \frac{(4\mu + 2\alpha_1 + \alpha_2 + \alpha_3) - (2\mu + \alpha_2 + \alpha_3)}{2} = \mu + \alpha_1.$$

In our illustration (Fig. 7.10), this corresponds to taking the sum over responses in the white and dark grey blocks, and subtracting the responses in the light grey blocks. Note that no information about treatment effects is contained in the block totals of an RCBD.

Analysis of Variance

Based on the linear model, the specification for the analysis of variance is `y~drug+Error(block)` for a random block factor, the same as for an RCBD. The resulting ANOVA table has again two error strata, and the block error stratum contains the inter-block information about the treatment effects. With two sums of squares, the analysis provides two different omnibus F-tests for the treatment factor.

	Df	Sum Sq	Mean Sq	F value	Pr(>F)
Error stratum: Batch					
Drug	2	14.82	7.41	3.42	7.88e-02
Residuals	9	19.53	2.17		
Error stratum: Within					
Drug	2	56.3	28.15	98.06	2.69e-07
Residuals	10	2.87	0.29		

The first F-test is based on the inter-block information about the treatment, and is in general (much) less powerful than the second F-test based on the intra-block information. Both F-tests can be combined to an overall F-test taking account of all information, but the inter-block F-test is typically ignored in practice and only the intra-block F-test is used and reported; this approximation leads to a (often small) loss in power.

The dependence of the block- and treatment factors must be considered for fixed block effects. The correct model specification contains the blocking factor *before* the treatment factor in the formula, and is `y~block+drug` for our example. This model adjusts treatments for blocks and the analysis is identical to an intra-block analysis for random block factors. The model `y~drug+block`, on the other hand, yields an entirely different ANOVA table and an incorrect F-test, as we discussed in Sect. 6.5.

Linear Mixed Model

The linear mixed model approach does not require error strata to cope with the two variance components, and provides a single estimate of the drug effects and treatment

Table 7.4 Contrast estimates and 95%-confidence intervals for BIBD based on intra-block estimates from classic analysis of variance (top) and linear mixed model (bottom)

Contrast	Estimate	se	df	LCL	UCL
ANOVA: aov(y~drug+Error(block))					
D1 – Placebo	4.28	0.31	10.00	3.59	4.97
D2 – Placebo	2.70	0.31	10.00	2.01	3.39
(D1+D2)/2 – Placebo	3.49	0.27	10.00	2.90	4.09
Mixed model: lmer(y~drug+(1|block))					
D1 – Placebo	4.22	0.31	10.81	3.54	4.90
D2 – Placebo	2.74	0.31	10.81	2.06	3.42
(D1+D2)/2 – Placebo	3.48	0.27	10.81	2.89	4.07

group means based on all available information. The resulting degrees of freedom are no longer integers and resulting F-tests and p-values can deviate from the classical analysis of variance. We specify the model as `y~drug+(1|block)`; this results in the ANOVA table

	Sum Sq	Mean Sq	NumDF	DenDF	F value	Pr(>F)
Drug	57.23	28.62	2	10.74	99.33	1.19e-07

We note that the sum of squares and the mean square estimates are slightly larger than for the `aov()` analysis, because the between-block information is taken into account. This provides more power and results in a slightly larger value of the F-statistic.

This is an example of a design in which the deliberate violation of complete balance still allows an analysis of variance, but where a mixed model analysis gives advantages both in the calculation but also the interpretation of the results. Since the BIBD is not fully balanced, the linear mixed model ANOVA table gives slightly different results when we approximate degrees of freedom with the more conservative Kenward–Roger method rather than the Satterthwaite method reported here.

7.3.4 Contrasts

Contrasts are defined exactly as for our previous designs, but their estimation is based only on the intra-block information if the estimated marginal means are calculated from the ANOVA model. Estimates and confidence intervals then differ between ANOVA and linear mixed model results, and the latter should be preferred. For our example, we calculate three contrasts comparing each drug, respectively, their average to placebo. The results are given in Table 7.4, and demonstrate the differences in degrees of freedom and precision between the two underlying models.

Estimation of all treatment contrasts is unbiased and contrast variances are free of the block variance in a BIBD. The variance of a contrast estimate is

$$\text{Var}(\hat{\Psi}(\mathbf{w})) = \frac{s}{\lambda k}\sigma_e^2 \sum_{i=1}^{k} w_i^2 \; .$$

where each treatment is assigned to r units. The corresponding contrast in an RCBD with the same number $n = r$ of replicates for each treatment (but lower number of blocks) has variance $\sigma_e^2 \sum_i w_i^2/n$.

We can interpret the term

$$\frac{\lambda k}{s} = r\frac{k}{s}\frac{s-1}{k-1} = n^*$$

as the "effective sample size" in each treatment group for a BIBD. It is smaller than the actual replication r since some information is lost as we cannot fully eliminate the block variances from the group mean estimates.

The relative efficiency of a BIBD compared to an RCBD with r blocks is

$$\text{RE}_\Psi(\text{BIBD}, \text{RCBD}) = \frac{r\frac{k}{s}\frac{s-1}{k-1}}{r} = \frac{n^*}{r} \; .$$

The fewer treatments fit into a single block, the lower the relative efficiency. For our example, $k = 3$ and $s = 2$ result in a relative efficiency of 3/4, and our BIBD requires about 33% more samples to achieve the same precision as an RCBD.

7.3.5 A Real-Life Example—Between-Plates Variability

As a real-world application, we consider an experiment to characterize and compare multiple antibody assays. Each assay only requires a small amount of patient serum, and in order to provide sufficient sample size for precisely estimating the sensitivity and specificity of each assay, several hundred patient samples were used. The samples were distributed over 10 96-well plates, and one concern was the reproducibility of results between plates, since each plate might be handled by a different technician and time between preparing the plate and pipetting each sample onto the assays might vary. For estimating between-plate variability, we would ideally create 10 aliquots from several patient sera, and assign one aliquot to each plate. However, the available sera only allowed at most five aliquots of sufficient volume. In this experiment, we are interested in contrasting the plates on the same patients, not the patients themselves. The 10 plates are then the treatment factor levels, and each patient is a block to which we assign plates.

A possible experimental design is a balanced incomplete block design with $k = 10$ plates assigned to blocks of size $s = 5$; this requires $b = 18$ patient sera with five aliquots each, a total of 90 aliquots. Each plate then contains serum samples of $r = 9$ patients, and each pair of plates co-occurs on $\lambda = 4$ patients. The design provides

Table 7.5 Balanced incomplete block design for estimating between-plate reproducibility for 10 plates A–J based on 18 patients samples, each patient sample in five aliquots

Aliquot	Patient																	
	1	2	3	4	5	6	7	8	9	10	11	12	13	14	15	16	17	18
1	D	F	E	G	F	J	E	H	A	I	A	B	A	H	H	G	H	I
2	G	H	F	F	H	D	B	C	F	F	E	D	J	J	I	H	E	B
3	A	J	I	J	E	I	F	B	C	J	D	F	E	B	A	C	I	E
4	B	A	C	I	J	B	H	J	H	D	J	I	G	I	G	D	A	A
5	C	D	G	A	G	E	D	G	B	C	C	G	B	C	D	E	C	F

the same precision for all pair-wise contrasts between plates and has about 89% efficiency. A possible layout is shown in Table 7.5, where letters correspond to the 10 plates.

7.3.6 Power Analysis and Sample Size

We conduct sample size determination and power analysis for a balanced incomplete block design in much the same fashion as for a randomized complete block design. The denominator degrees of freedom for the omnibus F-test are $\mathrm{df_{res}} = bk - b - k + 1$. The noncentrality parameter is

$$\lambda = k \cdot \left(r \cdot \frac{k\,s - 1}{s\,k - 1} \right) \cdot f^2 = k \cdot n^* \cdot f^2 \,,$$

which is again of the form "total sample size times effect size", but uses the effective sample size per group to adjust for the incompleteness.

7.3.7 Reference Designs

The *reference design* or *design for differential precision* is a variation of the BIBD that is useful when a main objective is comparing treatments to a control (or reference) group. One concrete application is a microarray study, where several conditions are to be tested, but only two dyes are available for each microarray. If all pair-wise contrasts between treatment groups are of equal interest, then a BIBD is a reasonable option, while we might prefer a reference design for comparing conditions against a common control as in Fig. 7.11A.

Another application of reference designs is the screening of several new treatments against a standard treatment. In this case, selected treatments might be com-

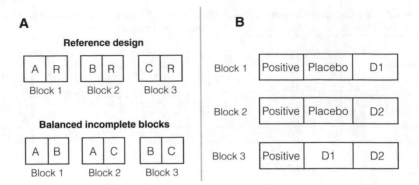

Fig. 7.11 **A** Reference design for three treatments A, B, C and a reference condition R (top) and BIBD without reference condition (bottom). **B** Reference design for three treatments and common positive control as reference condition with block size three

pared among each other in a subsequent experiment, and removal of unpromising candidates in the first round might reduce these later efforts.

Generating a simple reference design is straightforward: since each block of size s contains the reference condition, it has $s - 1$ remaining experimental units for accommodating a subset of the $k - 1$ non-control treatments. This translates into finding a BIBD for $k - 1$ treatments in blocks of size $s - 1$. The main analysis of this design is based on Dunnett-type contrasts, where each treatment level is compared to the reference level. The same considerations as for a CRD with Dunnett-type contrasts then apply, and we might think about using more than one unit per block for the reference condition to improve precision. A balance has then to be struck between improving the comparisons within each block by using more than one replicate of the control condition, and the need to use more blocks and introduce more uncertainty by having smaller block sizes available for the treatment conditions.

For our drug example, we might introduce a positive control condition in addition to the placebo group. A reference design with block size $s = 3$ then requires $b = 3$ blocks, with treatment assignments (positive, placebo, $D1$) for the first, (positive, placebo, $D2$) for the second, and (positive, $D1$, $D2$) for the third block; this is simply a BIBD with $s = 2$ and $k = 3$, augmented by adding the positive treatment to each block. A single replication is shown in Fig. 7.11B.

The same principle extends to constructing reference designs with more than one reference level. In our example, we might want to compare each drug to positive and placebo (negative) control groups, and we accommodate the remaining treatment factor levels in blocks of size $s - 2$. If the available block size is $s = 3$, then no remaining treatment levels co-occur in the same block. Comparisons of these levels to reference levels than profit from blocking, but comparisons between the remaining treatment levels do not.

7.4 Multiple Blocking Factors

7.4.1 Introduction

The unit structure of (G)RCBD and BIBD experimental designs consists of only two unit factors: the blocks and the experimental units nested in blocks. We now discuss unit structures with more than one blocking factor; these can be nested or crossed.

7.4.2 Nesting Blocks

Nesting of blocking factors allows us to replicate already blocked designs, to combine several properties for grouping, and to disentangle the variance components of different levels of grouping. It is also sometimes dictated by the logistics of the experiment. We consider only designs with two nested blocking factors, but our discussion directly extends to longer chains of nested factors.

We saw that we can replicate a completely randomized design by crossing its treatment structure with a factor to represent the replication; in our example, we used a CRD with three drugs in two laboratories, each laboratory providing one replicate of the CRD. The same idea allows replicating more complex designs: to replicate our RCBD for drug comparisons blocked by litter in two laboratories, for example, we again cross the treatment structure with a new factor *(Lab)* with two levels. Since each litter is unique to one laboratory, *(Litter)* from the original RCBD is nested in *(Lab)*, and both are crossed with **Drug**, since each drug is assigned to each litter in both laboratories. This yields the treatment and unit structures shown in Fig. 7.12A, B. The experiment design in Fig. 7.12C has two block-by-treatment interactions, and the three-way interaction is an interaction of a lab-specific litter with the treatment factor. This experiment cannot by analyzed using analysis of variance, because we cannot specify the block-by-treatment interactions; the linear mixed model is `y ~ drug + (1|lab/litter) + (1|lab:drug) + (1|lab:litter:drug)`; note that `(1|lab/litter)=(1|lab)+(1|lab:litter)`.

If we carefully select the blocking factors, we can assume these interactions to be negligible and arrive at the much simpler experiment design in Fig. 7.12D. Our experiment again uses 24 mice, with four litters per laboratory, and each litter is a block with one replicate per drug. The model specification is `y ~ drug + Error(lab/litter)` for an analysis of variance, and `y ~ drug + (1|lab/litter)` for a linear mixed model. As more than one litter is used per lab, the linear mixed model directly provides us with estimates of the between-lab and the between-litter (within lab) variance components. This gives insight into the sources of variation in this experiment and would allow us, for example, to conclude if the blocking by litter was successful on its own, and how discrepant values for the same drug are between different laboratories. Since the two blocks are nested,

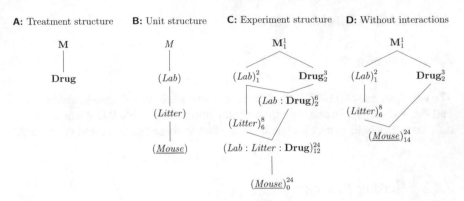

Fig. 7.12 Randomized complete block design for determining effect of two different drugs and placebo treatments on enzyme levels in mice, replicated in two laboratories. **A** Treatment structure. **B** Unit structure with three nested factors. **C** Full experiment structure with block-by-treatment interactions. **D** Experiment structure if these interactions are negligible

the omnibus F-test and contrasts for **Drug** are calculated within each litter and then averaged over litters within labs.

This type of design can be extended to an arbitrary number of nested blocks and we might use two labs, two cages per lab, and two litters per cage for our example. As long as each nested factor is replicated, we are able to estimate corresponding variance components. If a factor is not replicated (e.g., we use a single litter per lab), then there are no degrees of freedom for the nested blocking factor, and the effects of both blocking factors are completely confounded. Effectively, such a design uses a single blocking factor, where each level is a combination of lab and litter.

We should keep in mind that variance components are harder to estimate than averages, and estimates based on low replication are imprecise. This is not a problem if our goal is removal of variation from treatment contrasts, but sufficient replication is necessary if estimation of variance components is a primary goal of the experiment. Sample size determination for variance component estimation is covered in more specialized texts.

7.4.3 Crossing Blocks: Latin Squares

Nesting blocking factors essentially results in replication of (parts of) the design. In contrast, crossing blocking factors allows us to control several sources of variation simultaneously. The most prominent example is the *Latin square design*, which consists of two crossed blocking factors simultaneously crossed with a treatment factor, such that each treatment level occurs once in each level of the two blocking factors.

For example, we might be concerned about the effect of litters on our drug comparisons, but suspect that the position of the cage in the rack also affects the observations.

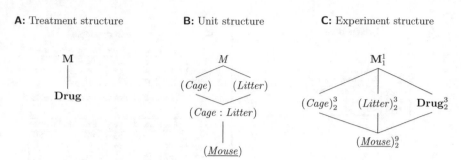

Fig. 7.13 Latin square with cage and litter as random row/column effects and three drug treatment levels.

The Latin square design removes both between-litter and between-cage variation from the drug comparisons. For three drugs, this design requires three litters of three mice each, and three cages. Crossing litters and cages results in one mouse per litter in each cage. The drugs are then randomized on the intersection of litters and cages (i.e., on mice) such that each drug occurs once in each cage and once in each litter.

The Hasse diagrams for this design are shown in Fig. 7.13. The crucial observation is that none of the potential block-by-block or block-by-treatment interactions can be estimated, since each combination of block levels or block-treatment combinations occurs only once. In particular, the intersection of a cage and a litter is a single mouse, and thus *(Cage:Litter)* is completely confounded with *(Mouse)* in this example.

An example layout of this design is shown in Fig. 7.14A, where the two blocking factors are given as rows and columns. Each treatment occurs exactly once per row and once per column and the Latin square design imposes two simultaneous constraints on the randomization of drugs on mice.

With all interactions negligible, data from a Latin square design are analyzed by an additive model

$$y_{ijk} = \mu + \alpha_i + c_j + r_k + e_{ijk} \, ,$$

where μ is the grand mean, α_i the treatment parameters, and c_j (resp. r_k) are the random column (resp. row) parameters. This model is again in direct correspondence to the experiment diagram and is specified as `y ~ drug + Error(cage+litter)`, respectively, `y ~ drug + (1|cage) + (1|litter)`.

We can interpret the Latin square design as a blocked RCBD: ignoring the cage effect, for example, the experiment structure is identical to an RCBD blocked by litter. Conversely, we have an RCBD blocked on cages when ignoring the litter effect. The remaining blocking factor blocks the whole RCBD. The requirement of having each treatment level once per cage and litter implies that the number of cages must equal the number of litters, and that both must also equal the number of treatments. These constraints pose some complications for the randomization of treatment levels to mice, and randomization is usually done by software; many older

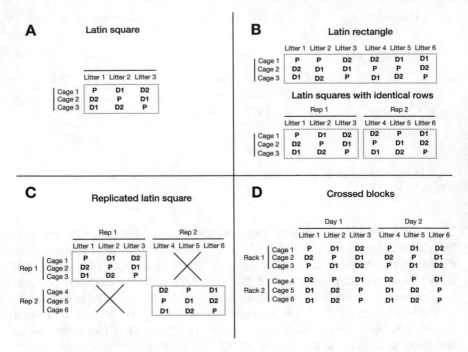

Fig. 7.14 **A** Latin square using cage and litter and three drug treatments. **B** Replication keeping the same cages, and using more litters without (top) and with (bottom) forming Latin squares. **C** Full replication with new rows and columns. **D** Two fully crossed blocks

books on experimental design also contain tables of Latin square designs with few treatment levels.

Replicating Latin Squares

A design with a single Latin square often lacks sufficient replication: for $k = 2, k = 3$, and $k = 4$ treatment levels, we only have 0, 2, and 6 residual degrees of freedom, respectively. We therefore consider several options for generating r replicates of a Latin square design. In each case, the Hasse diagram allows us to calculate the resulting error degrees of freedom and to specify an appropriate linear model.

First, we might consider replicating columns while keeping the rows, using rk column factor levels instead of k. The same logic applies to keeping columns and replicating rows, of course. The two experiments in Fig. 7.14B illustrate this design for a two-fold replication of the 3×3 Latin square, where we use six litters instead of three, but keep using the same three cages in both replicates. In the top part of the panel, we do not impose any new restrictions on the allocation of drugs and only require that each drug occurs the same number of times in each cage, and that each drug is used with each litter. In particular, the first three columns do not

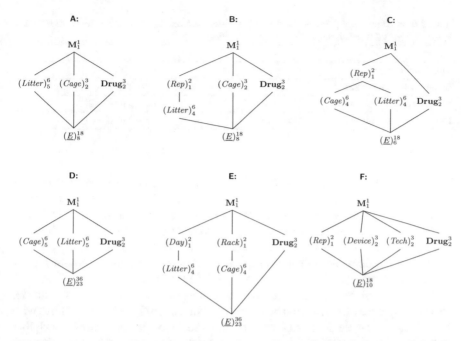

Fig. 7.15 Replication of latin square design. **A** Latin rectangle with six columns and three rows. **B** Two sets of three columns with identical rows. **C** New rows and columns in each replicate. **D** Fully crossed blocks, not a latin square. **E** Keeping both row and column levels with two independently randomized replicates. **F** Independent replication of rows and columns

form a Latin square in themselves. This design is called a *Latin rectangle*, and its experiment structure is shown in Fig. 7.15A with model specification `y ~ drug + Error(cage+litter)` or `y ~ drug + (1|cage)+(1|litter)`.

We can also insist that each replicate forms a proper Latin square in itself. That means we organize the columns in two groups of three as shown in the bottom of Fig. 7.14B. In the diagram in Fig. 7.15B, this is reflected by a new grouping factor *(Rep)* with two levels, in which the column factor *(Litter)* is nested. The model is specified as `y ~ drug + Error(cage+rep/litter)` or `y ~ drug + (1|cage)+(1|rep/litter)`.

By nesting both blocking factors in the replication, we find yet another useful design that uses different row and column factor levels in each replicate of the Latin square as shown in Fig. 7.14C. The corresponding diagram is Fig. 7.15C and yields the model `y ~ drug + Error(rep/(cage+litter))`, respectively, `y ~ drug + (1|rep) + (1|rep:cage) + (1|rep:litter)`.

We can also replicate both rows and columns (potentially with different numbers of replication) without restricting any subset of units to a Latin square. The experiment is shown in Fig. 7.14D and extends the Latin rectangle to rows and columns. Here, none of the replicates alone forms a Latin square, but all rows and columns have the same number of units for each treatment level. The diagram is shown in

Table 7.6 Noncentrality parameter for omnibus F-test and residual degrees of freedom for Latin square design and three strategies for r-fold replication

Design	NCP λ	df_{res}
Single latin square	$\lambda = k^2 \cdot f^2$	$(k-1)(k-2)$
Same rows, same columns	$\lambda = r \cdot k^2 \cdot f^2$	$(k-1)(r(k-1)-3)$
Same rows, new columns	$\lambda = r \cdot k^2 \cdot f^2$	$(k-1)(rk-2)$
New rows, new columns	$\lambda = r \cdot k^2 \cdot f^2$	$(k-1)(r(k-1)-1)$

Fig. 7.15D with model specification y ~ drug + Error(cage+litter) or y ~ drug + (1|cage)+(1|litter). Using this design but insisting on a Latin square in each of the four replicates yields the diagram in Fig. 7.15E with model y ~ drug + Error(rack/cage+day/litter) or y ~ drug + (1|rack/cage) + (1|day/litter). Given that we still have 23 residual degrees of freedom, we might also include the three remaining random effects (1|rack:day), (1|rack:litter), and (1|day:cage).

Finally, we might use the same row and column factor levels in each replicate. This is rarely the case for random block factors, but we could use our three drugs with three measurement devices and three laboratory technicians, for example, such that each technician measures one mouse for each treatment in each of the three devices. Essentially, this fully crosses the Latin square with a new blocking factor and leads to the model specification y ~ drug + Error(rep+device+tech) or y ~ drug + (1|rep) + (1|device) + (1|tech) for Fig. 7.15F.

Power Analysis and Sample size

The determination of sample size and power for a crossed-block design requires no new ideas. The noncentrality parameter λ is again the total sample size times the effect size f^2, and we find the correct degrees of freedom for the residuals from the experiment diagram.

For a single Latin square, the sample size is necessarily k^2 and increases to $r \cdot k^2$ if we use r replicates. The residual degrees of freedom then depend on the strategy for replication. The residual degrees of freedom and the noncentrality parameter are shown in Table 7.6 for a single Latin square and three replication strategies: using the same levels for row and column factors, using the same levels for either row or column factor and new levels for the other, and using new levels for both blocking factors. The numerator degrees of freedom are $k-1$ in each case.

Plant

Leaf height	I	II	III	IV	V	VI	VII
low	D	A	G	F	C	B	E
mid	A	C	D	G	B	E	F
high	B	E	C	A	F	G	D

Fig. 7.16 Experiment layout for Youden square. Seven plants are considered (I–VII) with seven different inoculation treatments (A–G). In each plant, three leaves at different heights (low/mid/high) are inoculated. Each inoculation occurs once at each height, and the assignment of treatments to plants forms a balanced incomplete block design

Youden Designs

The Latin square design requires identical number of levels for the row and column factors. We can use two blocking factors with a balanced incomplete block design to reduce the required number of levels for one of the two blocking factors. These designs are called *Youden squares* and only use a fraction of the treatment levels in each column (resp. row) and the full set of treatments in each row (resp. column). The idea was first proposed by Youden for studying inoculation of tobacco plants against the mosaic virus (Youden 1937), and his experiment layout is shown in Fig. 7.16.

In the experiment, seven plants were considered with seven different inoculation treatments, applied to individual leaves. It was suspected that the height of a plant's leaf influences the effect of the virus, and leaf height was used as a second blocking factor with three levels (low, middle, high). Crossing the two blocking factors leads to a 3×7 layout. The treatment levels are randomly allocated such that each treatment occurs once per leaf height (i.e., each inoculation occurs once in each column). The columns form a balanced incomplete block design of $k = 7$ treatments in $b = 7$ blocks of size $s = 3$, leading to $r = 3$ occurrences of each treatment, and $\lambda = 1$ blocks containing each pair of treatments, in accordance with the defining Eq. (7.3).

7.5 Notes and Summary

Notes

Different ways of replication lead to variants of the GRCBD and are presented in Addelman (1969) and Gates (1995). Variants of blocked designs with different block sizes are discussed in Pearce (1964), and the question of treating blocks as random or fixed in Dixon (2016). Evaluation and testing of blocks is reviewed in Samuels et al. (1991). Excellent discussions of interactions between different types of factors (e.g., treatment-treatment or treatment-classification) are given in Cox (1984) and de Gonzalez and Cox (2007).

Blocking designs are also important in animal experiments (Festing 2014; Lazic and Essioux 2013), and replicating pre-clinical experiments in at least two laboratories can greatly increase reproducibility (Karp 2018).

The idea of a Latin square can be extended to more than two blocking factors; with three factors, such designs are called *graeco-latin squares*.

Using R

Linear mixed models are covered in R by the lme4 package (Bates 2015) and its lmer() function, which we use exclusively in this book. This package does not provide *p*-values, which we can augment by additionally loading the lmerTest package (Kuznetsova et al. 2017). The linear mixed models are specified similarly to linear models, and random effects are introduced using the (X|G) construct that produces random effects for a factor X grouped by G. We only consider X=1, which produces random intercepts (or offsets) for each level of the factor G. Note that while crossed random factors R and C can be specified in aov() using Error(R+C), the equivalent notation (1|R+C) is *not* allowed in lmer(), and we use (1|R)+(1|C) instead. A comprehensive textbook on linear mixed models in R is Galecki and Burzykowski (2013); it captures many more uses of these models, such as analysis of longitudinal data.

Latin squares, balanced incomplete block designs and Youden designs are conveniently found and randomized by the functions design.lsd(), design.bib() and design.youden() in the agricolae package. For our example, design.bib(trt=c("Placebo", "D1", "D2"), k=2) yields our first BIBD with $k = 3$ and $s = 2$, and design.youden(trt=LETTERS[1:7], r=3) generates the Youden design of Fig. 7.16.

Contrast analysis is based on either aov() or lmer() for estimating the linear model, and estimated marginal means from emmeans(), where results from aov() are based exclusively on the intra-block information and can differ from those based on lmer().

Summary

Crossing a unit factor with the treatment structure leads to a blocked design, where each treatment occurs in each level of the blocking factor. This factor organizes the experimental units into groups, and treatment contrasts can be calculated within each group before averaging over groups. This effectively removes the variation captured by the blocking factor from any treatment comparisons. If experimental units are more similar within the same group than between groups, then this strategy can lead to substantial increase in precision and power, without increasing the sample size. The price we pay is slightly larger organizational effort to create the groups, randomize the treatments independently within each group, and to keep track of which experimental unit belongs to which group for the subsequent analysis.

The most common blocking design is the randomized complete block design, where each treatment occurs once per block. Its analysis requires the assumption of no block-by-treatment interaction, which the experimenter can ensure by suitable choice of the blocking factor. The efficiency of blocking is evaluated by appropriate effect sizes, such as the proportion of variation attributed to the blocking. It typically deteriorates if the block size becomes too large, since experimental units then become more heterogeneous. A balanced incomplete block design allows blocking of simple treatment structures if only a subset of treatments can be accommodated in each block.

Multiple blocking factors can be introduced in the unit structure. Nesting them enables estimation of their respective variance components, while crossing leads to row-column designs that control for two sources of variation simultaneously.

Blocked designs yield ANOVA results with multiple error strata, and only the lowest—within-block—stratum is typically used for analysis. Linear mixed models account for all information, and results might differ slightly from an ANOVA if the design is not fully balanced.

References

Addelman, S. (1969). "The generalized randomized block design". In: The American Statistician 23.4, pp. 35–36.

Bates, D. et al. (2015). "Fitting linear mixed-effects models using lme4". In: Journal of Statistical Software 67.1, pp. 1–48.

Cochran, W. G. and G. M. Cox (1957). Experimental Designs. John Wiley & Sons, Inc.

Cox, D. R. (1984). "Interaction". In: International Statistical Review 52, pp. 1–31.

de Gonzalez, A. and D. R. Cox (2007). "Interpretation of interaction: a review". In: The Annals of Applied Statistics 1.2, pp. 371–385.

Dixon, P. (2016). "Should Blocks Be Fixed or Random?" In: Conference on Applied Statistics in Agriculture.

Festing, M. F. W. (2014). "Randomized block experimental designs can increase the power and reproducibility of laboratory animal experiments". In: ILAR Journal 55.3, pp. 472–476.

Fisher, R. A. (1971). The Design of Experiments. 8th. Hafner Publishing Company, New York.

Galecki, A. and T. Burzykowski (2013). Linear mixed-effects models using R. Springer New York.

Gates, C. E. (1995). "What really is experimental error in block designs?" In: The American Statistician 49, pp. 362–363.

Karp, N. A. (2018). "Reproducible preclinical research-Is embracing variability the answer?" In: PLOS Biology 16.3, e2005413.

Kuznetsova, A., P. B. Brockhoff, and R. H. B. Christensen (2017). "lmerTest Package: Tests in Linear Mixed Effects Models". In: Journal Of Statistical Software 82.13, e1–e26.

Lazic, S. E. and L. Essioux (2013). "Improving basic and translational science by accounting for litter-to-litter variation in animal models". In: BMC Neuroscience 14.37, e1–e11.

Pearce, S. C. (1964). "Experimenting with Blocks of Natural Size". In: Biometrics 20.4, pp. 699–706.

Perrin, S. (2014). "Make mouse studies work". In: Nature 507, pp. 423–425.

Samuels, M. L., G. Casella, and G. P. McCabe (1991). "Interpreting Blocks and Random Factors". In: Journal of the American Statistical Association 86.415, pp. 798–808.

Watt, L. J. (1934). "Frequency Distribution of Litter Size in Mice". In: Journal of Mammalogy 15.3, pp. 185–189.

Youden, W. J. (1937). "Use of incomplete block replications in estimating tobacco-mosaic virus". In: Contributions from Boyce Thompson Institute 9, pp. 41–48.

Chapter 8
Split-Unit Designs

8.1 Introduction

In previous designs, we randomized all treatment factors on the same unit factor and these designs therefore have a single experimental unit factor. In some experimental setups, however, some treatment factors are more conveniently applied to groups of units while others can easily be allocated to individual units within groups.

For example, we might study the growth rate of a bacterium at different concentrations of glucose and different temperatures. Using 96-well plates for growing the bacteria, we can use a different amount of glucose for each well, but incubation restricts the whole plate to the same temperature. In other words, a well is the experimental unit for the glucose treatment while a plate is the experimental unit for the temperature treatment.

This kind of design is known as a *split-unit* (or *split-plot*) design, where (at least) two treatment factors (glucose concentration and temperature) are randomized on different nested unit factors (plates and wells nested in plates). The precision of a contrast estimate then depends on the treatment factors involved and their respective experimental units.

A related experimental design is the *criss-cross design* (commonly called *split-block* or *strip-plot*), where the two experimental unit factors are crossed rather than nested. This design naturally arises, e.g., when using a multi-channel pipette in a 96-well experiment: with one treatment per channel, all wells in a row of the plate contain the same treatment. Using different concentrations for a dilution series randomized over columns yields the second treatment and experimental unit since all wells in a column have the same dilution.

Both types of designs require care in the model specification to correctly reflect the relations between treatments and units. Otherwise, precision and power are overstated for some contrasts, resulting in deceptively low uncertainties and erroneous conclusions.

H.-M. Kaltenbach, *Statistical Design and Analysis of Biological Experiments*,
Statistics for Biology and Health, https://doi.org/10.1007/978-3-030-69641-2_8

8.2   Simple Split-Unit Design

We begin our discussion using two nested unit factors and two crossed treatment factors. A common application of the split-unit design is the accommodation of *hard-to-change* factors where applying a different level of one treatment factor is much more cumbersome than applying a different level of the other. To avoid frequent simultaneous changes of both levels, we keep the first treatment factor constant for a group of units, and randomize the second treatment factor within this group. This sacrifices precision and power for main effects of the first factor for the benefit of easier implementation. We call the first treatment factor the *whole-unit treatment*, and the group unit factor the *whole-unit*. We randomize the second treatment factor (the *sub-unit treatment*) on the nested unit factor (the *sub-unit*).

8.2.1   Experiment

We revisit our drug-diet example, with three drugs (placebo, $D1$, $D2$) combined with two diets (low fat, high fat) in an experiment with four mice per treatment, 24 mice in total, using enzyme level as our response.

In previous instances, we randomly assigned a drug-diet combination to each mouse (or each mouse in each block). To implement such an experiment, we have to individually apply the assigned drug to each mouse once at the beginning of the experiment. But we also have to feed each mouse its respective diet throughout the experiment; even if we hold several mice in one cage, we cannot apply the same diet to the whole cage, but have to individually feed each mouse within each cage.

A more practical implementation of the experiment uses eight cages with three mice, but while each mouse per cage is treated with a different drug, all mice in the same cage are fed the same diet. This makes each cage a block for the drugs, but the experimental unit for the diets. The experimental layout is shown in Fig. 8.1.

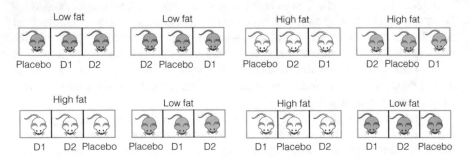

Fig. 8.1 Split-unit experiment with two diets randomized on cages of three mice, and three drugs randomized on mice within cages

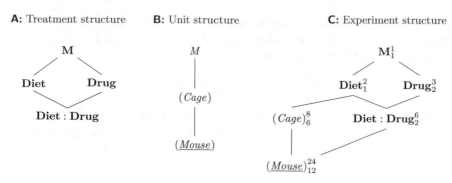

Fig. 8.2 Split-unit design with diets randomized on cages and drugs randomized on mice within cages. Cages are blocks for the drug treatment, but experimental units for the diet treatment

8.2.2 Hasse Diagram

The Hasse diagrams are constructed using our previous approaches and are shown in Fig. 8.2. The treatment structure is a 3×2 factorial with interaction. The unit structure consists of *(Mouse)* (quite literally) nested in *(Cage)*; since we measure one sample per mouse, *(Mouse)* is the response unit.

In contrast to previous designs, the two treatment factors now have *different* experimental units: we feed all mice in a cage the same diet, and **Diet** is randomized on *(Cage)*, while **Drug** is randomized on *(Mouse)*. Each level of the interaction **Diet:Drug** is a combination of a diet and a drug, and is randomly assigned to a mouse. As for most blocked designs, we assume that interactions between unit and treatment factors are negligible and do not include the factors *(Cage:***Drug***)* and *(Cage:***Diet:Drug***)*.

The experiment design diagram shows that *(Cage)* is a blocking factor for **Drug** and **Drug:Diet**; this removes the between-cage variation for contrasts of drug main effects and drug-diet interactions, but *not* for contrasts involving only **Diet**. Likewise, the presence of more than one mouse per cage looks like pseudo-replication for diet main effects, and increasing the number of mice per cage does not increase replication for **Diet**.

The F-test and contrasts for **Diet** are based on the degrees of freedom and the variation associated with *(Cage)*. Power and precision are therefore lower than for **Drug** and **Drug:Diet**, whose F-tests and contrasts are based on *(Mouse)*. The loss of precision for the whole-unit factor is the principal disadvantage of a split-unit design. For our purposes, the design is still successful: first, it achieves the desired simplified implementation of the experiment. Second, our main research question concerns the effects of the three drugs (the **Drug** main effect) and their modification by the diet (the **Drug:Diet** interaction). Both are based on the full replication and the lowest residual variance terms in the design. We are not interested in comparing only the diets themselves and our intended analysis is therefore largely unaffected by the comparatively low replication and precision for the **Diet** main effect.

The linear model for this design is

$$y_{ijk} = \mu + \alpha_i + \beta_j + (\alpha\beta)_{ij} + c_{jk} + e_{ijk} \,,$$

where $\alpha_i, \beta_j, (\alpha\beta)_{ij}$ are the drug and diet main effect parameters and the interaction parameters with $i = 1 \ldots 3$ and $j = 1 \ldots 2$. The random variables $c_{jk} \sim N(0, \sigma_c^2)$ are effects for the eight cages, and $e_{ijk} \sim N(0, \sigma_e^2)$ are the residuals within each cage with $k = 1 \ldots 4$.

8.2.3 Analysis of Variance

We derive the model specification directly from the experiment design diagram (Fig. 8.2C). All random factors are present in the unit structure, and the `Error()` term is therefore `Error(cage/mouse)` or simply `Error(cage)`. The fixed factors are all in the treatment structure, which is specified as `drug*diet`. The model specification is hence `y~drug*diet+Error(cage)`, leading to an ANOVA table with two error strata. We find each treatment factor exclusively in the error stratum of *its* experimental unit: **Diet** appears in the *(Cage)* error stratum and **Drug** and the interaction appear in the residual *(Mouse)* error stratum. The correct denominator for each F-test is found by starting from the corresponding treatment factor in the diagram, and following the edges downward until we find the first random factor: **Diet** is tested against the variation from cage to cage alone, and the F-test is based on one numerator and six denominator degrees of freedom. The resulting ANOVA table is

	Df	Sum Sq	Mean Sq	F value	Pr(>F)
Error stratum: Cage					
Diet	1	4.5	4.5	2.01	2.07e-01
Residuals	6	13.47	2.24		
Error stratum: Within					
Drug	2	47.11	23.55	26.17	4.21e-05
Drug:Diet	2	11.2	5.6	6.22	1.40e-02
Residuals	12	10.8	0.9		

Comparing the degrees of freedom in this table with those from the diagram confirms that our model specification corresponds to the design. Between-cage variation seems to be the dominant source of random variation in this experiment, and we are unable to detect any significant main effect for **Diet**. Both **Drug** and **Drug:Diet** are tested against the lower within-cage variation on twelve degrees of freedom, resulting in higher power.

8.2.4 Linear Mixed Model

An equivalent analysis using the linear mixed model uses the specification `y~drug*diet+(1|cage)`, where we directly find a between-cage variance of $\hat{\sigma}_c^2 = 0.45$, which is about half of the residual variance $\hat{\sigma}_e^2 = 0.9$, leading to an intra-class correlation of ICC=33%. The cages provide less efficient blocking than litters, but this is unproblematic since we introduced this factor to simplify the experiment implementation, and blocking for the drug effects is simply a welcome benefit.

The linear mixed model calculates sums of squares for **Diet** and *(Cage)* differently, but F-values and p-values are identical to the ANOVA:

	Sum Sq	Mean Sq	NumDF	DenDF	F value	Pr(>F)
Drug	47.11	23.55	2	12	26.17	4.21e-05
Diet	1.81	1.81	1	6	2.01	2.07e-01
Drug:Diet	11.2	5.6	2	12	6.22	1.40e-02

The interaction explains about 19% of the variation and its resulting F-test is statistically significant.

8.2.5 Contrast Analysis

We define and estimate linear contrasts based on a split-unit design in the same way as before, and can rely on estimated marginal means for providing the required treatment group means. Contrasts of drugs and of drug-diet interactions profit from higher replication and lower variance and are more precise than those comparing diets.

As an illustration, we first compare $D1$ and $D2$ to the placebo treatment separately under both diets and use a Dunnett-correction for multiple testing:

Contrast	Diet	Estimate	SE	df	LCL	UCL
D1—Placebo	low fat	2.66	0.67	12	0.97	4.35
D2—Placebo	low fat	3.22	0.67	12	1.53	4.91
D1—Placebo	high fat	4.07	0.67	12	2.38	5.77
D2—Placebo	high fat	1.30	0.67	12	-0.39	2.99

Precision decreases for contrasts that involve comparisons between diets, such as contrasting the placebo averages between the two diets:

Contrast	Estimate	SE	df	LCL	UCL
Placebo (high)–Placebo (low)	-0.7	0.82	14.74	-2.45	1.06

This contrast had the same precision as the four other contrasts in our previous designs, but has higher standard error and lower precision in this split-unit design.

8.2.6 Inadvertent Split-Unit Designs

The fact that several experimental unit factors are present requires particular care in setting up the analysis, and split-unit experiments are notorious for the many ways they can be incorrectly designed, analyzed, and interpreted. One problem is misspecification of the model. Starting from the Hasse diagram, this problem is easily avoided and the results can be checked by comparing the degrees of freedom between diagram and ANOVA table.

Another common problem is the *inadvertent split-unit design*, where an experiment is intended as, e.g., a completely randomized design but implemented as a split-unit design. Examples are numerous, particularly (but by no means exclusively) in the engineering literature on process optimization and quality control.

Inadvertent split-unit designs usually originate in the implementation phase, by deviating from the design table for a more convenient implementation. For example, a technician might realize that feeding mice by cage rather than individually simplifies the experiment, and create a split-unit design out of an anticipated CRD.

8.3 A Historical Example—Oat Varieties

In his classic paper 'Complex Experiments', Frank Yates reviews and expands the advances in statistical design of experiments since the 1920s (Yates 1935). The paper contains an experiment to investigate different varieties of oat using several levels of nitrogen as fertilizer, which we discuss as an additional example of a split-unit design with additional blocking.

The experiment is illustrated in Fig. 8.3A: three oat varieties 'Victory', 'Golden Rain', and 'Marvellous' (denoted $v_1 \ldots v_3$) are applied to plots of sufficient size. Meanwhile, four nitrogen levels $n_1 \ldots n_4$ are applied to smaller patches of land, denoted subplots (nested in plots). This yields a split-unit design with varieties randomized on plots, and nitrogen on subplots nested in plots.

A common problem in agricultural experimentation is the heterogeneity of the soil, exposure to sunlight, irrigation, and other factors, which add substantial variability between plots that are spatially more distant. In this example, the whole experiment is replicated in six blocks I ... VI, where each block consists of three neighboring plots, and varieties are independently randomized to plots within each block. This increases the replication to achieve precision of contrasts between varieties while simultaneously controlling for spatial heterogeneity over a large area. The design is therefore a split-unit design with a randomized complete block design on the whole-plot level.

The resulting 72 observations are shown in Fig. 8.3B, individually for each block. Block effects are clearly visible, and patterns are very similar between blocks, so assuming no block-by-treatment interaction seems reasonable. We also observe a

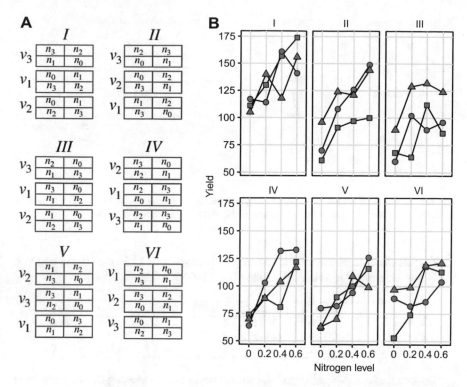

Fig. 8.3 **A** A Split-unit design with three oat varieties randomized on plots, four nitrogen amounts randomized on subplots within plots, and replication in six blocks. **B** Data shown separately for each block. Point: Golden Rain; triangle: Marvellous; square: Victory

pronounced trend of increasing yield with increasing nitrogen level, and this trend seems roughly linear. Differences between oat varieties are less obvious.

The Hasse diagrams are given in Fig. 8.4 and show the simple factorial treatment structure and the chain of nested unit factors combined into a fairly complex design, where the whole treatment structure is blocked, and the nitrogen and interaction treatment factors are blocked by the plots.

The original analysis in 1930 was of course done using an analysis of variance approach. Here, we analyze the experiment using a linear mixed model and derive

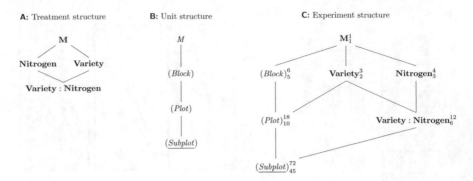

Fig. 8.4 Hasse diagram for Yates' oat variety and nitrogen example with two treatment factors randomized on plots, respectively, subplots in plots, and replication in six blocks

the specification `Variety*Nitrogen+(1|Block)+(1|Plot:Block)` from the Hasse diagram. This yields the following ANOVA table:

	Sum Sq	Mean Sq	NumDF	DenDF	F value	Pr(>F)
Variety	526.06	263.03	2	10	1.49	2.72e-01
Nitrogen	20020.5	6673.5	3	45	37.69	2.46e-12
Variety:Nitrogen	321.75	53.62	6	45	0.3	9.32e-01

The small and non-significant interaction shows that increasing the nitrogen level has roughly the same effect on yield for all three oat varieties. In addition, differences between oat varieties are also small with average yields between 80 and 175 and differences all less than 10, and not significant. The nitrogen level, on the other hand, shows a large and highly significant effect, and higher levels give more yield.

We further quantify these findings by estimating corresponding contrasts and their confidence intervals. First, we compare the varieties within each nitrogen level (Table 8.1). In each case, *Marvellous* provides higher yield than both *Golden Rain* and *Victory*, and *Golden Rain* gives higher yield than *Victory*: the varieties have a clear order, which is stable over all nitrogen levels. As the confidence intervals show, however, none of the differences are significant, and the precision of estimates is fairly low.

For quantifying the dose-response relationship between nitrogen level and yield, we estimate the nitrogen main effect contrasts independently within each oat variety. We use a polynomial contrast for `Nitrogen`, which provides information about linear, quadratic, and cubic components of a dose-response. The results are shown in Table 8.2.

For each variety, we find a substantial linear upward trend. Since both quadratic and cubic terms are small and not significant, we can ignore all potential curvature in the trends and arrive at an easy to interpret result: the yield increases proportionally with increases in nitrogen level. We already determined that the average and nitrogen-level-specific yields are almost identical between varieties. The current contrasts

Table 8.1 Comparing the three oat varieties within each level of nitrogen

Contrast	Estimate	se	df	LCL	UCL
Nitrogen: 0.0					
Golden Rain—Marvellous	−6.67	9.71	30.23	−30.61	17.27
Golden Rain—Victory	8.50	9.71	30.23	−15.44	32.44
Marvellous—Victory	15.17	9.71	30.23	−8.77	39.11
Nitrogen: 0.2					
Golden Rain—Marvellous	−10.00	9.71	30.23	−33.94	13.94
Golden Rain—Victory	8.83	9.71	30.23	−15.11	32.77
Marvellous—Victory	18.83	9.71	30.23	−5.11	42.77
Nitrogen: 0.4					
Golden Rain—Marvellous	−2.50	9.71	30.23	−26.44	21.44
Golden Rain—Victory	3.83	9.71	30.23	−20.11	27.77
Marvellous—Victory	6.33	9.71	30.23	−17.61	30.27
Nitrogen: 0.6					
Golden Rain—Marvellous	−2.00	9.71	30.23	−25.94	21.94
Golden Rain—Victory	6.33	9.71	30.23	−17.61	30.27
Marvellous—Victory	8.33	9.71	30.23	−15.61	32.27

Table 8.2 Orthogonal contrasts for nitrogen levels within each oat variety show linear dose-response relation

Contrast	Estimate	se	df	t value	P(>|t|)
Golden Rain					
Linear	150.67	24.30	45	6.20	0.00
Quadratic	−8.33	10.87	45	−0.77	0.45
Cubic	−3.67	24.30	45	−0.15	0.88
Marvellous					
Linear	129.17	24.30	45	5.32	0.00
Quadratic	−12.17	10.87	45	−1.12	0.27
Cubic	14.17	24.30	45	0.58	0.56
Victory					
Linear	162.17	24.30	45	6.67	0.00
Quadratic	−10.50	10.87	45	−0.97	0.34
Cubic	−16.50	24.30	45	−0.68	0.50

additionally show that the estimates of the three linear components are all within roughly one standard error of each other, demonstrating a comparable dose-response relation for all three varieties. This of course agrees with the previous result that there is no variety-by-nitrogen interaction.

8.4 Variations and Related Designs

The split-unit design turns out to be quite ubiquitous in experimental work. We briefly discuss several variations of this design idea to explore some additional uses: accommodating an additional factor in an already existing design, using more than two nested units and randomizing a treatment factor on each level of nesting, and using crossed rather than nested experimental units for the treatment allocation. We also introduce the simplest case of cross-over designs, where two treatments are used in sequence on the same experimental unit. Finally, longitudinal experimental designs involving a (usually temporal) order of treatments and measurements such as for comparing a response before and after application of a treatment are sometimes considered and analyzed as split-unit designs.

8.4.1 Accommodating an Additional Factor

We turned our previous drug-diet example into a split-unit design by grouping mice into cages and using the new grouping factor as experimental unit for the diets. This creates a whole-plot factor 'above' the original experimental unit. Similar to our discussion of choosing a blocking factor for an RCBD, we can alternatively sub-divide the original experimental unit further to create a sub-plot factor 'below'.

To illustrate this idea, we consider the following situation: we start from our original drug-diet design with factorial treatment structure randomized on mice (a CRD). Previously, we also considered comparing two sample preparation kits from vendors A and B based on the enzyme level measurements. Since we already have our drug-diet experiment planned, we would like to 'squeeze' the comparison of the two kits into that experiment without jeopardizing our main objective of estimating contrasts of the drug-diet treatments.

The idea is simple: we draw two samples per mouse and randomly assign either kit A or kit B to each sample. The resulting experiment structure is shown in Fig. 8.5A, and we recognize it as a split-unit design. Here, the whole-plot unit *(Mouse)* is combined with a factorial treatment structure, and the sub-plot unit *(Sample)* is nested in *(Mouse)* to compare levels of **Vendor**. The resulting treatment structure is a $3 \times 2 \times 2$ factorial, where we removed all interactions involving **Vendor** under the assumption that these are negligible. The original drug-diet experiment is then unaffected by this augmentation of the design: even if vendor B's kit is worse, we

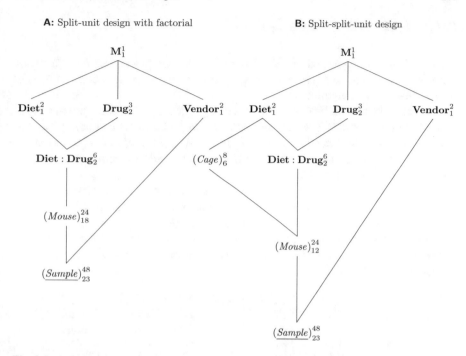

Fig. 8.5 A Split-unit design with diets and drugs completely randomized on mice as a CRD and vendor randomized on samples. **B** Same treatment structure with split-split-unit design

still have the full data for vendor A; simply removing the B data yields the data for the originally anticipated design.

We use the linear mixed model framework for estimating the corresponding model with specification `y~drug*diet+vendor+(1|mouse)` and estimating the difference between the two vendors.

Contrast	Estimate	se	df	LCL	UCL
Vendor A–Vendor B	−0.4	0.18	23	−0.77	−0.02

This contrast is estimated very precisely with 23 residual degrees of freedom, the same as for a randomized complete block design with 24 mice as blocks and two samples per mouse and no other treatment factors. It has much higher precision than the drug or diet comparisons, because each mouse provides a block for **Vendor** to compare the two kits *within each mouse*.

8.4.2 Split-Split-Unit Designs

By introducing three nested unit factors and randomizing one treatment factor on each, we arrive at a *split-split-unit* design. Further extensions to arbitrary levels of nested factors are straightforward.

For example, we combine the split-unit design for drugs and diets with a comparison of the two vendors. The new design is shown in Fig. 8.5B and uses *(Cage)* as experimental unit for the *hard-to-change* factor **Diet**, *(Mouse)* in *(Cage)* as experimental unit for **Drug**, and *(Sample)* in *(Mouse)* in *(Cage)* to accommodate **Vendor** as an additional treatment factor.

From the diagram, we find one random intercept for each cage, leading to a random effect term (1|cage), one random intercept for each mouse within a cage, with (1|cage:mouse), and the omitted (1|cage:mouse:sample). A linear mixed model that ignores all interactions of **Vendor** with other factors is therefore specified as y~drug*diet+vendor+(1|cage)+(1|cage:mouse) and yields the ANOVA table

	Sum Sq	Mean Sq	NumDF	DenDF	F value	Pr(>F)
Drug	26.3	13.15	2	12	33.44	1.24e-05
Diet	1.02	1.02	1	6	2.6	1.58e-01
Vendor	1.89	1.89	1	23	4.81	3.87e-02
Drug:Diet	4.51	2.26	2	12	5.74	1.78e-02

The results are very similar to our split-unit design without the additional **Vendor** treatment. Interactions involving **Vendor** can of course be introduced, and lead to a more complex analysis and interpretation of results.

8.4.3 Criss-Cross or Split-Block Designs

In contrast to the split-unit design, we cross the two unit factors in a *criss-cross design* and combine this unit structure with a factorial treatment structure. The simplest instance of a criss-cross design is a row-column design with a rows and b columns, where a treatment factor with a levels is randomized on the rows, and a crossed treatment factor with b levels is randomized on columns. This treatment structure is an $a \times b$ factorial, but each treatment factor has its own experimental unit. In contrast to a split-unit design, the interaction of the two treatment factors does not share its experimental unit with any of the main effect factors. For a two-way factorial treatment structure, the criss-cross design therefore has three experimental units and such a design needs to be replicated several times to arrive at suitable residual degrees of freedom for all experimental unit factors. Usually, the rows and columns are independently replicated, and randomization is done independently for each replicate of the row-column criss-cross design.

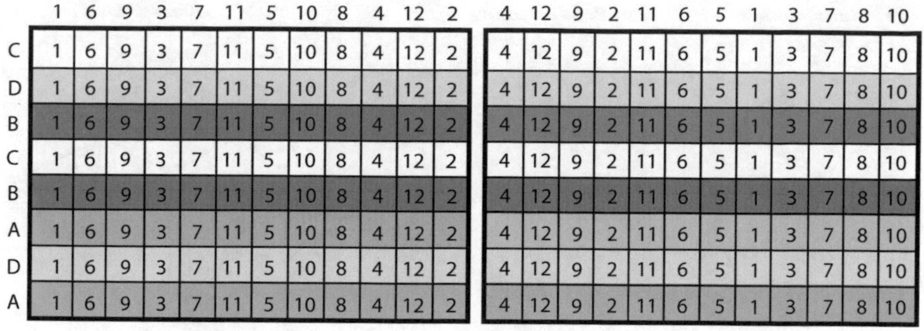

Fig. 8.6 Criss-cross experiment layout: two replicates of four drugs (background shade) randomized on rows, dilutions (numbers) randomized on columns. Two replicate plates shown, randomization of rows kept constant while dilutions are randomized independently

Example: Multi-Channel Pipetting

The criss-cross design rather naturally arises in experiments on 96-well plates when using multi-channel pipettes; common multi-channel pipettes offer eight channels such that eight consecutive wells can be handled simultaneously.

This setup is advantageous in assays based on dilution series, where up to eight different conditions are subjected to twelve dilutions each. A typical response is the optical density in each well, for example. Using one pipette channel for each condition allows randomization of the conditions on the rows of each plate, but the same condition is then assigned to all wells in the same row. Similarly, the dilution steps can be randomized on columns, but each of the eight rows then has a fixed dilution. This arrangement leads to a criss-cross design with conditions randomized on rows by randomly assigning them to the channels of the pipette at the beginning of the experiment, and dilutions randomized on columns.

The plate layouts in Fig. 8.6 show a version of this strategy for comparing the effect of four drugs on bacterial growth in twelve glucose concentrations in the growth medium. Two channels are randomly assigned to each drug and each glucose level is used on one full column; we use two plates to provide higher replication. For easier implementation, the assignment of drugs to pipette channels is only randomized once, and then kept identical for both plates. The glucose levels are randomized independently to columns for each plate. This provides an interesting variant of the criss-cross design.

The Hasse diagrams for this example are shown in Fig. 8.7. The treatment structure is a simple two-way factorial design of drug and glucose. In the unit structure, columns are nested in plates since randomization is independent between plates, but rows are crossed with plates since any row in the first plate is identified with the corresponding row in the second plate. We omitted several interaction factors that we assume negligible for brevity, but the experiment structure is already rather complex.

From the experiment diagram, we derive the model specifications `y ~ drug * glucose + Error(plate/col + row)` for an ANOVA and `y ~ drug`

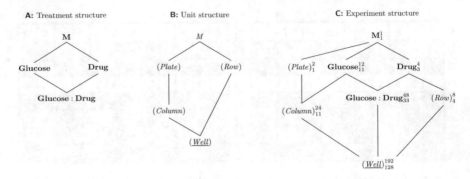

Fig. 8.7 Criss-cross design arising from use of multi-channel pipette. Four drugs are tested with 12 glucose concentrations on each plate, two plates provide replication. Use of 8-channel pipette allows two replicates of each drug; random assignment of drug to channel is kept constant for both plates, but assignment of glucose concentration to columns is randomized independently

`* glucose + (1|plate/col) + (1|row)` for a linear mixed model. The classical ANOVA table has four error strata: one for *(Plate)*, one for *(Column)* in *(Plate)* containing the **Glucose** main effect, one for *(Row)* containing the **Drug** main effect, and the residual error stratum containing the **Glucose:Drug** interaction. The linear mixed model provides direct estimation of the four variances in this model, and its ANOVA table is

	Sum Sq	Mean Sq	NumDF	DenDF	F value	Pr(>F)
Drug	30.35	10.12	3	4	6.32	5.35e-02
Glucose	38.7	3.52	11	11	2.2	1.04e-01
Drug:Glucose	46.82	1.42	33	128	0.89	6.47e-01

The denominator of the three treatment F-tests corresponds to the closest random factor below its treatment factor. With several random factors crossed and nested, traditional ANOVA and linear mixed model results differ; we would prefer the latter.

8.4.4 Cross-Over Designs

A useful design for increasing precision and power is the *cross-over design*, where different treatments are assigned in sequence to the same experimental unit. As a basic example, we consider an experiment for determining the effect of the low- and high-fat diet (with no drug treatment) on the enzyme levels. We use six mice, which we split into two groups: we feed the mice in the first group on the low-fat diet for some time, and then switch them to the high-fat diet. In the second group, we reverse the order and feed first the high-fat diet, and then the low-fat diet. This is a *two-period two-treatment cross-over* design. The experiment is illustrated in Fig. 8.8 for three mice per group.

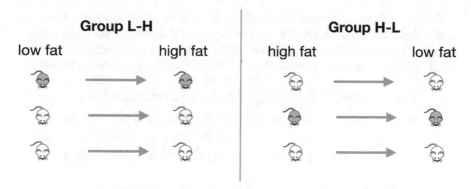

Fig. 8.8 Cross-over experiment with two diets assigned in one of two orders

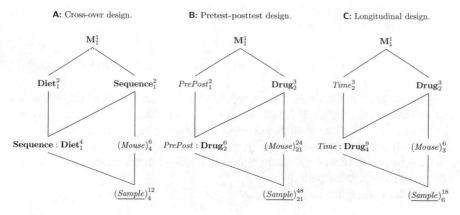

Fig. 8.9 **A** Cross-over design uses two diet treatments sequentially on same mouse to provide within-mouse contrasts. **B** Pretest-posttest design with measurement before and after application of treatment to consider mouse-specific baseline response values. **C** Longitudinal repeated measures design to allow multiple measurements of same mouse at different time-points.

Before each diet treatment, we feed all mice with a standard diet. This should allow the enzyme level to reset to 'normal', such that the first diet does not affect observations with the second diet. The observations are taken after several days on the respective diet, with one observation per mouse per diet.

The experiment diagram is shown in Fig. 8.9A. The treatment factor **Sequence** denotes the group: each mouse is assigned to either the low-high (L-H) sequence of diets, or the high-low (H-L) sequence. The sequence is crossed with the second treatment factor **Diet**, since each diet occurs in each sequence. Each level of the interaction **Sequence:Diet** corresponds to the application of one diet at a specific part in each sequence (the *period*). Each mouse is randomly assigned to one sequence, so *(Mouse)* is the experimental unit for **Sequence**. Each sample corresponds to a combination of a period and a diet, and is the experimental unit for **Diet** and the interaction.

It is instructive to consider the effects associated with **Sequence**, **Diet**, and **Sequence:Diet**. Since each factor has one degree of freedom, these effects are simple differences. A commonly used model for a two-period two-treatment cross-over design yields the following expected values for the four period-diet combinations:

$$\mu_{LH1} = \mu + \alpha_L + \pi_1 \qquad \text{(low-high, first observation)}$$
$$\mu_{LH2} = \mu + \alpha_H + \gamma_L + \pi_2 \qquad \text{(low-high, second observation)}$$
$$\mu_{HL1} = \mu + \alpha_H + \pi_1 \qquad \text{(high-low, first observation)}$$
$$\mu_{HL2} = \mu + \alpha_L + \gamma_H + \pi_2 \qquad \text{(high-low, second observation) ,}$$

where μ is the grand mean, α_i are the effects of the low and high fat diet, π_j is the effect of period j, and γ_k are the residual *carry-over* effects from the previous diet not eliminated by the *washout period* between diets.

The **Sequence** main effect is

$$\frac{1}{2}(\mu_{LH1} + \mu_{LH2}) - \frac{1}{2}(\mu_{HL1} + \mu_{HL2}) = \frac{1}{2}(\gamma_L - \gamma_H) ,$$

with associated hypothesis $H_0 : \gamma_L = \gamma_H$ that the two carry-over effects are equal (but not necessarily zero!). This test essentially asks if there is a difference between the two orders in which the diets are applied. If both carry-over effects are equal, then no difference exists since then $\gamma_L = \gamma_H = \gamma$, and we can merge γ with the period effect π_2 (all observations are higher or lower by the same amount in the first compared to the second period).

The **Diet** main effect is

$$\frac{1}{2}(\mu_{LH1} + \mu_{HL2}) - \frac{1}{2}(\mu_{HL1} + \mu_{LH2}) = \alpha_L - \alpha_H - \frac{1}{2}(\gamma_L - \gamma_H) ,$$

and is biased if the two carry-over effects are not equal. Note that we can in principle estimate and test the bias from the **Sequence** main effect, but that this effect has lowest replication in the design, and low precision and power. In the case of unequal carry-over effects, one often restricts the analysis to data from the first period alone, and estimates the treatment effect via $(\mu_{LH1} - \mu_{HL1})/2$.

The **Sequence:Diet** interaction effect is

$$\frac{1}{2}(\mu_{LH1} - \mu_{LH2}) - \frac{1}{2}(\mu_{HL1} - \mu_{HL2}) = \pi_1 - \pi_2 - \frac{1}{2}(\gamma_L + \gamma_H) ,$$

and is biased whenever there are—even equal—carry-over effects.

Cross-over designs form an important class of designs and the two-period two-treatment design is only the simplest instance. It does not allow estimation of the carry-over effects, which is a major weakness in practice where carry-over can often be suspected and the experiment should provide information about its magnitude. Better variants of the cross-over design that allow explicit estimation of the carry-over should therefore be preferred whenever feasible. One variant also uses two periods, but includes the two combinations H-H and L-L in addition to H-L and L-H. Carry-over can then be estimated by comparing the H-H to the L-H observations, for example. Another variant extends the design to three periods, with treatment sequences including H-H-L and L-H-L, for example, such that one treatment is observed twice in each sequence. The references in Sect. 8.5 provide more in-depth coverage of different cross-over designs and associated analyses.

8.4.5 Pretest-Posttest Designs

Another common technique to increase precision is the *pretest-posttest design*, where the response variable is measured once before and once after the treatment is applied. This provides a simple way for adjusting the treatment response by a subject-specific baseline, and the difference between response after treatment and baseline is then considered as the relevant quantity for estimating the treatment effect. We study a simple example of a pretest-posttest design with a single treatment factor, but the ideas readily extend to factorial treatment structures as well.

We consider our experiment for comparing three drugs, and use the baseline enzyme levels of each mouse in conjunction with the enzyme level after adminis-tration of the drug. The experiment diagram in Fig. 8.9B illustrates this design. It contains three unit factors, *(Sample)* nested in *(Mouse)*, since we take two samples from each mouse, one before, one after the drug administration, and the fixed unit factor *PrePost*, which designates if a sample was taken before or after treatment is administered. The design contains **Drug** as the only treatment factor, crossed with *PrePost*. We introduce their interaction as a third factor into the design. Since both samples belong to the same mouse, and a drug is applied to a mouse after the baseline measurement is taken, *(Mouse)* is the experimental unit for **Drug**. The correspond-ing models are `y ~ prepost * drug + Error(mouse)` for `aov()`, and `y ~ prepost * drug + (1|mouse)` for `lmer()`. Note that while this design looks very similar to a cross-over design, only *one* treatment is assigned to each mouse and sample.

Because the unit factor *PrePost* is fixed, there are three F-tests: the *PrePost* main effect compares the average response over all drugs before and after administration. We expect that the measured enzyme levels are not systematically different between the three drug groups before applying the treatment. Thus, a small and non-significant pre-post main effect either indicates that the before and after responses are identical for all drugs; none of the drugs has any discernible effect. Or it might be that one drug increases the enzyme level, and another drug decreases it, and the two effects cancel out.

The F-test of the **Drug** main effect tests if the average enzyme levels are identical for all three drugs, when before and after measurements are lumped together. The denominator mean squares for this test stem from the between-mouse variation. This test is the least powerful, but also the least interesting.

Of greatest interest is usually the *PrePost*:**Drug** interaction, which shows how different the changes of enzyme levels are between drugs from baseline to post-treatment measurement. This is the drug effect corrected for the baseline measurement. We can replicate the corresponding F-test as follows: for each mouse i, calculate the difference $\Delta_i = y_{i,\text{post}} - y_{i,\text{pre}}$ of the post-treatment response and the pre-treatment response. This 'adjusts' the response to the treatment for the baseline value. Now, we perform a one-way ANOVA with **Drug** as the treatment factor, and Δ_i as the response variable. The resulting F-ratio and p-value are identical to the *PrePost*:**Drug** test.

8.4.6   Longitudinal Designs

Split-unit designs are sometimes still used for *repeated measures* and *longitudinal* designs, in which multiple response variables are measured for the same experimental unit, respectively, the same response variable is measured at multiple occasions for the same experimental unit. Both designs thus have a more complex response structure than the classical approach can handle.

An example of a longitudinal design is shown in Fig. 8.9C, where three drugs are randomized on two mice each, and each mouse is then measured at three time-points. In this design, we randomize **Drug** on *(Mouse)*, and the fixed unit factor *Time* groups the samples from each mouse. We can then relate observations from the same mouse to each other to analyze the temporal profile of each mouse. The advantage of the longitudinal design is that observations can be contrasted within each mouse, and the between-mouse variation is removed from such contrasts.

The main caveat of this approach is the crude approximation of the complex longitudinal response structure by a fixed block factor *Time*. This assumes that any pair of time-points has the same correlation, while observations closer in time often tend to have stronger correlations than those further apart. This caveat does not apply to the pretest-posttest designs, where only two time-points are considered.

8.5   Notes and Summary

Notes

Insightful accounts on split-unit designs are Federer (1975) and Box (1996), and a gentle introduction is given in Kowalski and Potcner (2003). Recent developments in split-unit designs are reviewed in Jones and Nachtsheim (2009). Analysis of split-unit designs with more complex whole-unit and sub-unit treatment designs are discussed in Goos and Gilmour (2012), and power analysis in Kanji and Liu (1984). Increasing availability of liquid-handling robots renewed interest in split-unit and criss-cross designs for microplate-based experiments (Buzas et al. 2011).

Different variants of cross-over designs are discussed in detail in Johnson (2010) and Senn (1994), and a tutorial framing the analysis of cross-over designs into linear contrasts is Shuster (2017). A standard text for cross-over designs in clinical trials is Senn (2002), and applications to bioassays were highlighted already in the 1950s (Finney 1956).

An extensive treatment of pretest-posttest designs is Bonate (2000), and an illustrative example comparing two surgical techniques with several different analysis techniques is given in Brogan and Kutner (1980).

The use of split-unit-like designs for longitudinal experiments usually requires analysis techniques with *non-sphericity corrections* that account for unequal correlation between pairs of observations in the same group (Abdi 2010; Huynh and Feldt 1976; Greenhouse and Geisser 1959). By now, more appropriate models—including more complex variants of the linear mixed model—are available and should be preferred (Fitzmaurice et al. 2011; Diggle et al. 2013).

Using R

The function `design.split()` generates and randomizes split-unit designs; the option `design=` allows a CRD, RCBD, or Latin squares design for the whole-unit stratum. Criss-cross designs are generated and randomized by the function `design.strip()`. Both functions are from the `agricolae` package.

The biggest difficulty in using split-unit designs in R is usually the model specification; this problem is largely alleviated when using Hasse diagrams, from which the specification is directly derived.

Summary

Split-unit designs offer flexibility in the implementation of an experiment when some treatment factors (sometimes called hard-to-change factors) are more conveniently randomized on groups of units, while other treatment factors can be randomized on the units themselves. In laboratory work, split-unit designs are often required when different experimental conditions are examined on a 96-well plate, for example, but factors like temperature or shaking frequency apply to the plate as a whole.

With two experimental unit factors, sub-unit treatments have higher replication and are more precisely estimated than whole-unit treatment effects; importantly, however, their interaction also profits from high replication and high precision.

Several common designs such as cross-over designs can be seen as split-unit designs in disguise. Longitudinal designs, including the pretest-posttest design, are sometimes treated as split-unit designs; as an important conceptual difference, however, the whole-unit treatment is replaced by a fixed unit factor to capture the more complex response structure.

References

Abdi, H. (2010). "The Greenhouse-Geisser correction". In: Encyclopedia of Research Design. Ed. by Neil Salkind. SAGE Publications, Inc.

Bonate, P. L. (2000). Analysis of Pretest-Posttest Designs. Chapman & Hall/CRC.

Box, G. E. P. (1996). "Quality quandaries: Split plot experiments". In: Quality Engineering 8.3, pp. 515–520.

Brogan, D. R. and M. H. Kutner (1980). "Comparative Analyses of Pretest-Posttest Research Designs". In: The American Statistician 34.4, pp. 229–232.

Buzas, J. S., C. G. Wager, and D. M. Lansky (2011). "Split-Plot Designs for Robotic Serial Dilution Assays". In: Biometrics 67.4, pp. 1189–1196.

Diggle, P. et al. (2013). Analysis of Longitudinal Data. 2nd. Oxford University Press.

Federer, W. T. (1975). "The misunderstood split plot". In: Applied Statistics. Ed. by R. P. Gupta. Amsterdam: North-Holland, pp. 9–39.

Finney, D. J. (1956). "Cross-Over Designs in Bioassay". In: Proceedings of the Royal Society B: Biological Sciences 145.918, pp. 42–61.

Fitzmaurice, G. M., N. M. Laird, and J. H. Ware (2011). Applied longitudinal analysis. John Wiley & Sons, Inc.

Goos, P. and S. G. Gilmour (2012). "A general strategy for analyzing data from split-plot and multistratum experimental designs". In: Technometrics 54.4, pp. 340–354.

Greenhouse, S. W. and S. Geisser (1959). "On methods in the analysis of profile data". In: Psychometrika 24.2, pp. 95–112.

Huynh, H. and L. S. Feldt (1976). "Estimation of the Box correction for degrees of freedom from sample data in randomized block and split-plot designs". In: Journal Of Educational Statistics 1.1, pp. 69–82.

Johnson, D. E. (2010). "Crossover experiments". In: Wiley Interdisciplinary Reviews: Computational Statistics 2.5, pp. 620–625.

Jones, B. and C. J. Nachtsheim (2009). "Split-plot designs: what, why, and how". In: Journal of Quality Technology 41.4, pp. 340–361.

Kanji, G. K. and C. K. Liu (1984). "Power Aspects of Split-Plot Designs". In: Journal of the Royal Statistical Society. Series D 33.3, pp. 301–311.

Kowalski, S. M. and K. Potcner (2003). "How to recognize a split-plot experiment". In: Quality Progress 36.11, pp. 60–66.

Senn, S. J. (1994). "The AB/BA crossover: past, present and future?" In: Statistical Methods in Medical Research 3, pp. 303–324.

Senn, S. J. (2002). Cross-over trials in clinical research. 2nd. Wiley, New York, p. 364.

Shuster, J. J. (2017). "Linear combinations come alive in crossover designs". In: Statistics in Medicine 36.24, pp. 3910–3918.

Yates, F. (1935). "Complex Experiments". In: Journal of the Royal Statistical Society 2.2, pp. 181–247.

Chapter 9
Many Treatment Factors: Fractional Factorial Designs

9.1 Introduction

Factorial treatment designs are necessary for estimating factor interactions and offer additional advantages (Chap. 6). However, their implementation is challenging if we consider many factors or factors with many levels, because the number of treatments might then require prohibitive experiment sizes. Large factorial experiments also pose problems for blocking, if reasonable block sizes that ensure homogeneity of the experimental material within a block are smaller than the number of treatment level combinations.

For example, a factorial treatment structure with five factors of two levels each has $2^5 = 32$ treatment combinations. An experiment with 32 experimental units then has no residual degrees of freedom, but two full replicates of this design already require 64 experimental units. If each factor has three levels, the number of treatment combinations increases drastically to $3^5 = 243$.

On the other hand, we might justify the assumption of *effect sparsity*: high-order interactions are often negligible, especially if interactions of lower orders already have small effect sizes. The key observation for reducing the experiment size is that a large portion of model parameters relate to higher-order interactions: in a 2^5-factorial, there are 32 model parameters: one grand mean, five main effects, 10 two-way interactions, 10 three-way interactions, five four-way interactions, and one five-way interaction. The number of higher-order interactions and their parameters grows fast with increasing number of factors as shown in Table 9.1 for factorials with two factor levels and 3 to 7 factors.

If we ignore three-way and higher interactions in the example, we remove 16 parameters from the model equation and only require 16 observations for estimating the remaining model parameters; this is known as a *half-fraction* of the 2^5-factorial. Of course, the ignored interactions do not simply vanish, but their effects are now *confounded* with those of lower-order interactions or main effects. The question then arises: which 16 out of the 32 possible treatment combinations should we consider such that no effect of interest is confounded with a non-negligible effect?

© Springer Nature Switzerland AG 2021
H.-M. Kaltenbach, *Statistical Design and Analysis of Biological Experiments*,
Statistics for Biology and Health, https://doi.org/10.1007/978-3-030-69641-2_9

Table 9.1 Number of parameters for effects of different order in 2^k-designs

Factorial	Effect order							
	0	1	2	3	4	5	6	7
$k = 3$	1	3	3	1				
$k = 4$	1	4	6	4	1			
$k = 5$	1	5	10	10	5	1		
$k = 6$	1	6	15	20	15	6	1	
$k = 7$	1	7	21	35	35	21	7	1

In this chapter, we discuss the construction and analysis of *fractional replications* of 2^k-factorial designs where all k treatment factors have two levels. This restriction is often sufficient for practical experiments with many factors, where interest focuses on identifying relevant factors and low-order interactions. We first consider generic factors which we call **A**, **B** and so forth, and denote their two levels generically as *low* (or -1) and *high* (or $+1$).

We also extend the idea of fractional replication to deliberately confound some effects with blocks. This allows us to implement a 2^5-factorial in blocks of size 16, for example. By altering the confounding between pairs of blocks, we can still recover all effects, albeit with reduced precision.

9.2 Aliasing in the 2^3-Factorial

9.2.1 Introduction

We begin our discussion with the simple example of a 2^3-factorial treatment structure in a completely randomized design. We denote the treatment factors as **A**, **B**, and **C** and their levels as A, B, and C with values -1 and $+1$, generically called the *low* and *high* level, respectively. Recall that main effects and interactions (of any order) all have one degree of freedom in a 2^k-factorial; hence, we can encode the two independent levels of an interaction as -1 and $+1$. We define the level by multiplying the levels of the constituent factors: for $A = -1$, $B = +1$, $C = -1$, the level of **A:B** is $AB = A \cdot B = -1$ and the level of **A:B:C** is $ABC = A \cdot B \cdot C = +1$.

It is also convenient to use an additional shorthand notation for a treatment combination, where we use a character string containing the lower-case letter of a treatment factor if it is present on its high level, and no letter if it is present on its low level. For example, we write abc if **A**, **B**, **C** are on level $+1$, and all potential other factors are on the low level -1, and ac if **A** and **C** are on the high level, and **B** on its low level. We denote a treatment combination with all factors on their low level by (1). For a 2^3-factorial, the eight different treatments are then (1), a, b, c, ab, ac, bc, and abc.

Table 9.2 Eight treatment level combinations for 2^3-factorial with corresponding level of interactions and shorthand notation

Carbon	Nitrogen	Vitamin	A	B	C	AB	AC	BC	ABC	Shorthand
Glc	low	Mix 1	-1	-1	-1	$+1$	$+1$	$+1$	-1	(1)
Glc	low	Mix 2	-1	-1	$+1$	$+1$	-1	-1	$+1$	c
Glc	high	Mix 1	-1	$+1$	-1	-1	$+1$	-1	$+1$	b
Glc	high	Mix 2	-1	$+1$	$+1$	-1	-1	$+1$	-1	bc
Fru	low	Mix 1	$+1$	-1	-1	-1	-1	$+1$	$+1$	a
Fru	low	Mix 2	$+1$	-1	$+1$	-1	$+1$	-1	-1	ac
Fru	high	Mix 1	$+1$	$+1$	-1	$+1$	-1	-1	-1	ab
Fru	high	Mix 2	$+1$	$+1$	$+1$	$+1$	$+1$	$+1$	$+1$	abc

For example, testing compositions for growth media with factors **Carbon** with levels Glc (glucose) and Fru (fructose), **Nitrogen** with levels low and high, and **Vitamin** with levels Mix 1 and Mix 2 leads to a 2^3-factorial with the 8 possible treatment combinations shown in Table 9.2.

9.2.2 Effect Estimates

In a 2^k-factorial, we estimate main effects and interactions as simple contrasts by subtracting the sum of responses of all observations with the corresponding factors on the 'low' level -1 from those with the factors on the 'high' level $+1$. For our example, we estimate the main effect of **Carbon** (or generically **A**) by subtracting from each observation with fructose ('high') the corresponding observation with glucose ('low') with nitrogen and vitamin on the same level, and averaging:

$$\text{A main effect} = \frac{1}{4}\left((a - (1)) + (ab - b) + (ac - c) + (abc - bc)\right) .$$

This is equivalent to calculating the **Carbon** main effect by averaging the difference between the sum of observations with glucose ('low') and the sum of observations with fructose ('high'). In terms of Table 9.2, this amounts to adding all those observations for which $A = +1$, namely a, ab, ac, abc and subtracting the sum of all observations for which $A = -1$, namely $(1), b, c, bc$. This yields

$$\text{A main effect} = \frac{1}{4}\left(\underbrace{(a + ab + ac + abc)}_{A=+1} - \underbrace{((1) + b + c + bc)}_{A=-1}\right),$$

which we see is simply the previous calculation with terms grouped differently.

A two-way interaction is a difference of differences and we find the interaction of **B** with **C** by first finding the difference between them for **A** on the low level and for **A** on the high level:

$$\underbrace{\frac{1}{2}\left((abc-ab)-(ac-a)\right)}_{\text{if } A=+1} \quad \text{and} \quad \underbrace{\frac{1}{2}\left((bc-b)-(c-(1))\right)}_{\text{if } A=-1}.$$

The interaction effect is then the average of these two estimates

$$\text{B:C interaction} = \frac{1}{2}\left(\frac{1}{2}\left((abc-ab)-(ac-a)\right)+\frac{1}{2}\left((bc-b)-(c-(1))\right)\right)$$

$$= \frac{1}{4}\left(\underbrace{(abc+bc+a+(1))}_{BC=+1}-\underbrace{(ab+ac+b+c)}_{BC=-1}\right).$$

This value is equivalently found by taking the difference between observations with $BC = +1$ (the interaction at its 'high' level) and $BC = -1$ (the interaction at its 'low' level) and averaging. The other interaction effects are estimated by contrasting the corresponding observations for $AB = \pm 1$, $AC = \pm 1$, and $ABC = \pm 1$, respectively.

9.2.3   Reduction to Four Treatment Combinations

We are interested in reducing the size of the experiment and for reasons that will become clear shortly, we choose a design based on measuring the response for four out of the eight treatment combinations. This will only allow estimation of four parameters in the linear model, and exactly which parameters can be estimated depends on the treatments chosen. The question then is which four treatment combinations should we select?

We investigate three specific choices to get a better understanding of the consequences for effect estimation. The designs are illustrated in Fig. 9.1, where treatment level combinations form a cube with eight vertices, from which four are selected in each case.

First, we arbitrarily select the four treatment combinations (1), a, b, ac (Fig. 9.1A). With this choice, *none* of the main effects or interaction effects can be estimated using all four observations. For example, an estimate of the **A** main effect involves $a - (1)$, $ab - b$, $ac - c$, and $abc - bc$, but only $a - (1)$ is available in this experiment. Compared to a factorial experiment in four runs, this choice of treatment combinations thus allows using only one-half of the available data for estimating this effect. If we follow the above logic and contrast the observations with **A** at the high level with those with **A** at the low level, thereby using all data, then the main effect is estimated as $(ac + a) - (b + (1))$ and leads to a biased and incorrect estimate of the main

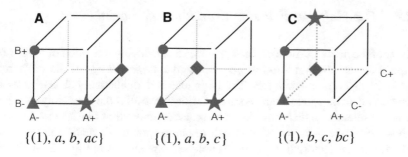

$$\{(1), a, b, ac\} \qquad \{(1), a, b, c\} \qquad \{(1), b, c, bc\}$$

Fig. 9.1 Subsets of a 2^3-factorial. **A** Arbitrary choice of treatment combinations leads to problems in estimating any effects properly. **B** One variable at a time (OVAT) design. **C** Keeping one factor at a constant level confounds this factor with the grand mean and creates a 2^2-factorial of the remaining factors

effect, since the other factors are at 'incompatible' levels. Similar problems arise for **B** and **C** main effects, where only $b - (1)$, respectively, $ac - a$ are available. None of the interactions can be estimated from these data and we are left with a very unsatisfactory muddle of biased estimates.

Next, we try to be more systematic and select the four treatment combinations $(1), a, b, c$ (Fig. 9.1B) where all factors occur on low and high levels. Again, main effect estimates are based on half of the data for each factor, but their calculation is now simpler: $a - (1), b - (1)$, and $c - (1)$, respectively. Each estimate involves the same level (1) and only two of four observations are used. This design resembles a one variable at a time experiment, where effects can be estimated individually for each factor, but no estimates of interactions are available. All advantages of a factorial treatment design are then lost.

Finally, we select the four treatment combinations $(1), b, c, bc$ with **A** on the low level (Fig. 9.1C). This design is effectively a 2^2-factorial with treatment factors **B** and **C** and allows estimation of their main effects and their interaction, but no information is available on any effects involving the treatment factor **A**. For example, we estimate the **B** main effect as $(bc + b) - (c + (1))$ using all data, and the **B:C** interaction as $(bc - b) - (c - (1))$. If we look more closely into Table 9.2, we find a simple confounding structure: the level of **B** is always the negative of **A:B**. In other words, the two effects are completely confounded in this design, and $(bc + b) - (c + (1))$ is in fact an estimate of the *difference* of the **B** main effect and the **A:B** interaction. Similarly, **C** is the negative of **A:C**, and **B:C** is the negative of **A:B:C**. Finally, the grand mean is confounded with the **A** main effect; this makes sense since any estimate of the overall average is based only on the 'low' level of **A**.

9.2.4 The Half-Replicate or Fractional Factorial

Neither of the previous three choices provides a convincing reduction of the factorial design. We now discuss a fourth possibility, the *half-replicate* of the 2^3-factorial, called a 2^{3-1}-*fractional factorial*. The main idea is to deliberately alias a high-order interaction with the grand mean. For a 2^3-factorial, we alias the three-way interaction **A:B:C** by selecting either those four treatment combinations that have $ABC = -1$ or those that have $ABC = +1$. We call the corresponding equation the *generator* of the fractional factorial; the two possible sets are shown in Fig. 9.2. With either choice, we find three more effect aliases by consulting Table 9.2. For example, using $ABC = +1$ as our generator yields the four treatment combinations a, b, c, abc and we find that **A** is completely confounded with **B:C**, **B** with **A:C**, and **C** with **A:B**.

In this design, any estimate thus corresponds to the sum of two effects. For example, $(a + abc) - (b + c)$ estimates the sum of **A** and **B:C**: first, the main effect of **A** is found as the difference of the runs a and abc with **A** on its high level, and the runs b and c with **A** on its low level: $(a + abc) - (b + c)$. Second, we contrast runs with **B:C** on the high level (a and abc) with those with **B:C** on its low level (b and c) for estimating the **B:C** interaction effect, which is again $(a + abc) - (b + c)$.

The fractional factorial based on this generator hence deliberately aliases each main effect with a two-way interaction, and the grand mean with the three-way interaction. Each estimate is then the sum of the two aliased effects. Moreover, we note that by pooling the treatment combinations over levels of one of the three factors, we create three different 2^2-factorials for the two remaining factors as seen in Fig. 9.2. For example, ignoring the level of **C** leads to the full factorial in **A** and **B**. This is a consequence of the aliasing, as **C** is completely confounded with **A:B**.

The confounding of different effects can be described by the *alias sets*, where each set contains the effects that cannot be distinguished. For the generator $ABC = +1$, the alias sets are

$$\{1, ABC\}, \quad \{A, BC\}, \quad \{B, AC\}, \quad \{C, AB\},$$

and for the generator $ABC = -1$, the alias sets are

$$\{1, -ABC\}, \quad \{A, -BC\}, \quad \{B, -AC\}, \quad \{C, -AB\}.$$

Estimation of the **A** main effect, for example, is only possible if the **B:C** interaction is zero in line with our previous observations. A more detailed discussion of confounding in terms of the parameters of the underlying linear model is given in Sect. 9.9.

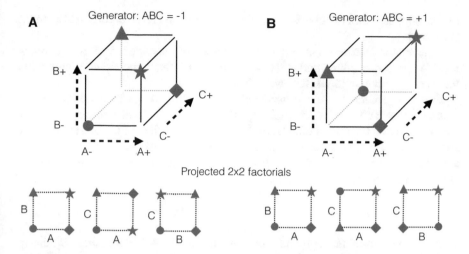

Fig. 9.2 The two half-replicates of a 2^3-factorial with three-way interaction and grand mean confounded. Any projection of the design to two factors yields a full 2^2-factorial design and main effects are confounded with two-way interactions. **A** Design based on low level of three-way interaction; **B** Complementary design based on high level

9.3 Aliasing in the 2^k-Factorial

The half-replicate of a 2^3-factorial still does not provide an entirely convincing example for the usefulness of fractional factorial designs due to the complete confounding of main effects and two-way interactions, both of which are typically of great interest. With more factors in the treatment structure, however, we are able to alias interactions of higher order and confound low-order interactions of interest with high-order interactions that we might assume negligible.

9.3.1 Using Generators

The generator or *generating equation* provides a convenient way for constructing fractional factorial designs. The generator is a word written by concatenating the factor letters, such that AB denotes a two-way interaction, and our previous example ABC is a three-way interaction; the special 'word' 1 denotes the grand mean. A generator is then a formal equation that identifies two words and enforces the equality of the corresponding treatment combinations. In our 2^{3-1} design, the generator

$$ABC = +1$$

selects all those rows in Table 9.2 for which the relation is true and **A:B:C** is on the high level.

A generator determines the effect confounding of the experiment: the generator itself is one confounding, and $ABC = +1$ describes the complete confounding of the three-way interaction **A:B:C** with the grand mean.

From the generator, we can derive all other confoundings by simple algebraic manipulation. By formally 'multiplying' the generator with an arbitrary word, we find a new relation between effects. In this manipulation, the multiplication with the letter $+1$ leaves the equation unaltered, multiplication with -1 inverses signs, and a product of two identical letters yields $+1$. For example, multiplying our generator $ABC = +1$ with the word B yields

$$ABC \cdot B = (+1) \cdot B \iff AC = B .$$

In other words, the **B** main effect is confounded with the **A:C** interaction. Similarly, we find $AB = C$ and $BC = A$ as two further confounding relations by multiplying the generator with C and A, respectively.

Further trials with manipulating the generator show that we can obtain no additional relations. For example, multiplying $ABC = +1$ with the word AB yields $C = AB$ again, and multiplying this relation with C yields $C \cdot C = AB \cdot C \iff +1 = ABC$, the original generator. This means that indeed, we have fully confounded four pairs of effects and no others. In general, a generator for a 2^k-factorial produces $2^k/2 = 2^{k-1}$ alias relations between factors, so we have a direct way to check if we found all. In our example, $2^3/2 = 4$, so our relations $ABC = +1$, $AB = C$, $AC = B$, and $BC = A$ cover all existing aliases.

This property also means that we arrive at exactly the same set of alias relations, no matter which of them we choose as our generator. For example, instead of $ABC = +1$, we might choose $A = BC$; this selects the same set of rows and implies the same set of confounding relations. Usually, we use a generator that aliases the highest-order interaction with the grand mean and yields the least severe confounding.

Generators provide a systematic way for aliasing that results in interpretable effect estimates with known confoundings. A generator selects one-half of the possible treatment combinations, and this is the reason why we set out to choose four rows in our first example.

We briefly note that our first and second choices in Sect. 9.2.3 are not based on a generator, leaving us with a complex partial confounding of effects. In contrast, our third choice selected all treatments with **A** on the low level and does have a generator, namely

$$A = -1 .$$

Algebraic manipulation then shows that this design implies the additional three alias relations $AB = -B$, $AC = -C$, and $ABC = -BC$. In other words, any effect involving the factor **A** is confounded with another effect not involving that factor, which we easily verify from Table 9.2.

9.3.2 Half-Replicates

Generators and their algebraic manipulation provide an efficient way for finding the confoundings in higher-order factorials, where looking at the corresponding table of treatment combinations quickly becomes unfeasible. As we can see from the algebra, the most useful generator is always confounding the grand mean with the highest-order interaction.

For four factors, this generator is $ABCD = +1$ and we expect that there are $2^4/2 = 8$ relations in total. Multiplying with any letter reveals that main effects are then confounded with three-way interactions, such as $ABCD = +1 \iff BCD = A$ after multiplying with A, and similarly $B = ACD$, $C = ABD$, and $D = ABC$. Moreover, by multiplication with two-letter words we find that all two-way interactions are confounded with other two-way interactions, namely via the three relations $AB = CD$, $AC = BD$, and $AD = BC$. We thus found eight relations and can be sure that there are no others.

The resulting confounding is already an improvement over fractions of the 2^3-factorial, especially if we can make the argument that three-way interactions can be neglected and we thus have direct estimates of all main effects. If we find a significant and large two-way interaction—**A:B**, say—then we cannot distinguish if it is **A:B**, its alias **C:D**, or a combination of the two that produces the effect. Subject-matter considerations might be available to separate these possibilities. If not, there is at least a clear goal for a subsequent experiment to disentangle the two interaction effects.

Things improve further for five factors and the generator $ABCDE = +1$ which reduces the number of treatment combinations from $2^5 = 32$ to $2^{5-1} = 16$. Now, main effects are confounded with four-way interactions, and two-way interactions are confounded with three-way interactions. Invoking the principle of effect sparsity and neglecting the three- and four-way interactions yields main effects and two-way interactions as the estimated parameters.

Main effects and two-way interactions are confounded with interactions of order four or higher for factorials with six factors and more, and we can often assume that these interactions are negligible.

9.4 A Real-Life Example—Yeast Medium Composition

As a concrete example of a fractional factorial treatment design, we discuss an experiment conducted during the sequential optimization of a yeast growth medium, which we discuss in more detail in Chap. 10. For now, we concentrate on determining the individual and combined effects of five medium ingredients—glucose **Glc**, two different nitrogen sources **N1** (monosodium glutamate) and **N2** (an amino acid mixture), and two vitamin sources **Vit1** and **Vit2**—on the resulting number of yeast cells. Different combinations of concentrations of these ingredients are tested on a 48-well plate, and the growth curve is recorded for each well by measuring the optical density over time. We use the increase in optical density (ΔOD) between onset of growth and flattening of the growth curve at the diauxic shift as a rough but sufficient approximation for increase in number of cells.

9.4.1 Experimental Design

To determine how the five medium components influence the growth of the yeast culture, we use the composition of a standard medium as a reference point, and simultaneously alter the concentrations of the five components. For this, we select two concentrations per component, one lower, the other higher than the standard, and consider these as two levels for each of five treatment factors. The treatment structure is then a 2^5-factorial and would in principle allow estimation of the main effects and all two-, three-, four-, and five-factor interactions when we use all 32 possible combinations. However, a single replicate requires two-thirds of a 48-well plate and this is undesirable because we would like sufficient replication and also be able to compare several yeast strains in the same plate. Both requirements can be accommodated by using a half-replicate of the 2^5-factorial with 16 treatment combinations, such that three independent experiments fit on a single plate.

A generator $ABCDE = +1$ confounds the main effects with four-way interactions, which we consider negligible for this experiment. Still, two-way interactions are confounded with three-way interactions, and in the first implementation we assume that three-way interactions are much smaller than two-way interactions. We can then interpret main effect estimates directly, and assume that estimates of parameters involving two-way interactions have only small contributions from the corresponding three-way interactions. The design is shown in Table 9.3.

We use two replicates of this design for adequate sample size, requiring 32 wells in total. This could also accommodate the full 2^5-factorial, but we would then have no replication for estimating the residual variance. Moreover, our duplicate of the same design enables inspection of reproducibility of measurements and detection of errors and aberrant observations. The observed increase in optical density is shown in Table 9.3 with columns 'OD 1' and 'OD 2' for the two replicates.

Clearly, the medium composition has a huge impact on the resulting growth, ranging from a minimum of close to zero to a maximum of 216.6. The original medium has an average 'growth' of ΔOD $\approx$ 80, and this experiment already reveals a condition with approximately 2.7-fold increase. We also see that observations with **N2** at the low level are abnormally low in the first replicate and we remove these eight values from further analysis.[1]

---

[1] It later transpired that the low level of **N2** had zero concentration in the first, but a low, non-zero concentration in the second replicate.

Table 9.3 Treatment combinations for half-replicate of 2^5-factorial design for determining yeast growth medium composition. Last two columns show responses for two replicates, observations in italics result from experimental error and are removed from analysis

Glucose	Nitrogen 1	Nitrogen 2	Vitamin 1	Vitamin 2	OD 1	OD 2
−1	−1	−1	−1	1	*1.7*	35.68
1	−1	−1	−1	−1	*0.1*	67.88
−1	1	−1	−1	−1	*1.5*	27.08
1	1	−1	−1	1	*0*	80.12
−1	−1	1	−1	−1	120.2	143.39
1	−1	1	−1	1	140.3	116.30
−1	1	1	−1	1	181	216.65
1	1	1	−1	−1	40	47.48
−1	−1	−1	1	−1	*5.8*	41.35
1	−1	−1	1	1	*1.4*	5.70
−1	1	−1	1	1	*1.5*	84.87
1	1	−1	1	−1	*0.6*	8.93
−1	−1	1	1	1	106.4	117.48
1	−1	1	1	−1	90.9	104.46
−1	1	1	1	−1	129.1	157.82
1	1	1	1	1	131.5	143.33

9.4.2   Analysis

Our fractional factorial design has five treatment factors and several interaction factors, and we initially use an analysis of variance to determine which of the medium components has an appreciable effect on growth, and how the components interact. The model `Growth~(Glc+N1+N2+Vit1+Vit2)^2` yields the ANOVA table

	Df	Sum Sq	Mean Sq	F value	Pr(>F)
Glc	1	6147.53	6147.53	26.49	8.77e-04
N1	1	1038.3	1038.3	4.47	6.73e-02
N2	1	34297.69	34297.69	147.82	1.94e-06
Vit1	1	369.94	369.94	1.59	2.42e-01
Vit2	1	6039.65	6039.65	26.03	9.28e-04
Glc:N1	1	3906.52	3906.52	16.84	3.42e-03
Glc:N2	1	1939.07	1939.07	8.36	2.02e-02
Glc:Vit1	1	264.76	264.76	1.14	3.17e-01
Glc:Vit2	1	753.29	753.29	3.25	1.09e-01
N1:N2	1	0.93	0.93	0	9.51e-01
N1:Vit1	1	1449.59	1449.59	6.25	3.70e-02
N1:Vit2	1	9357.9	9357.9	40.33	2.20e-04
N2:Vit1	1	277.86	277.86	1.2	3.06e-01
N2:Vit2	1	811.42	811.42	3.5	9.84e-02
Vit1:Vit2	1	1279.62	1279.62	5.51	4.68e-02
Residuals	8	1856.21	232.03		

The specification expands to `Growth~Glc+N1+...+Glc:N1+...+Vit1:Vit2` and describes a model with main effects and all two-way interactions. The full model is `Growth~Glc*N1*N2*Vit1*Vit2` and additionally includes three-, four-, and five-way interactions. However, only half of its parameters can be estimated. Since we deliberately confounded effects in our fractional factorial treatment structure, we know which parameters are aliased, and can select one member of each alias set in the model specification.

We find several substantial effects in this analysis, with **N2** the main contributor followed by **Glc** and **Vit2**. Even though **N1** has no significant main effect, it appears in several significant interactions; this also holds to a lesser degree for **Vit1**. Several pronounced interactions demonstrate that optimizing individual components will not be a fruitful strategy, and we need to simultaneously change multiple factors to maximize the growth. This information can only be acquired by using a factorial design.

We do not discuss the necessary subsequent analyses of contrasts and effect sizes for the sake of brevity; they work exactly as for smaller factorial designs.

9.4.3   Alternative Analysis of Single Replicate

If only the single replicate is available, then we have to reduce the model to free up degrees of freedom from parameter estimation to estimate the residual variance (cf. Sect. 6.4.2). If subject-matter knowledge is available to decide which factors can be safely removed without missing important effects, then a single replicate can be successfully analyzed. For example, knowing that the two nitrogen sources and the two vitamin components do not interact, we might specify the model `Growth~(Glc+N1+N2+Vit1+Vit2)^2 - N1:N2 - Vit1:Vit2` that removes the two corresponding interactions while keeping the three remaining ones. This strategy is somewhat unsatisfactory, since we now still only have two residual degrees of freedom and correspondingly low precision and power, and we cannot test if removal of the factors was really justified. Without good subject-matter knowledge, this strategy can give very misleading results if significant and large effects are removed from the analysis.

9.5   Multiple Aliasing

For higher-order factorials starting with the 2^5-factorials, useful designs are also available for higher than one-half fractions, such as quarter-replicates that would require only 8 of the 32 treatment combinations in a 2^5-factorial. These designs are constructed by using more than one generator, and combined aliasing leads to more complex confounding of effects.

For example, a quarter-fractional requires two generators: one generator to specify one-half of the treatment combinations, and a second generator to specify one-half of those. Both generators introduce their own aliases which we determine using

Table 9.4 Quarter-fractionals of 2^5-factorial. Left: Generators $ABCDE = +1$ and $BCDE = +1$ confound main effect of A with grand mean. Right: Generators $ABD = +1$ and $ACE = +1$ confound main effects with two-way interactions

A	B	C	D	E	ABCDE	BCDE	A	B	C	D	E	ABD	ACE
+1	−1	−1	−1	−1	+1	+1	+1	−1	−1	−1	−1	+1	+1
+1	+1	+1	−1	−1	+1	+1	−1	+1	+1	−1	−1	+1	+1
+1	+1	−1	+1	−1	+1	+1	+1	+1	−1	+1	−1	+1	+1
+1	−1	+1	+1	−1	+1	+1	−1	−1	+1	+1	−1	+1	+1
+1	+1	−1	−1	+1	+1	+1	−1	+1	−1	−1	+1	+1	+1
+1	−1	+1	−1	+1	+1	+1	+1	−1	+1	−1	+1	+1	+1
+1	−1	−1	+1	+1	+1	+1	−1	−1	−1	+1	+1	+1	+1
+1	+1	+1	+1	+1	+1	+1	+1	+1	+1	+1	+1	+1	+1

the generator algebra. In addition, multiplying the two generators introduces further aliases through the *generalized interaction*.

9.5.1 A Generic 2^{5-2}-Fractional Factorial

As a first example, we construct a quarter-replicate of a 2^5-factorial, also called a 2^{5-2}-fractional factorial. Our first idea is probably to use the five-way interaction for defining the first set of aliases, and one of the four-way interactions for defining the second set. For example, we might choose the two generators G_1 and G_2 as

$$G_1 : ABCDE = +1 \quad \text{and} \quad G_2 : BCDE = +1 .$$

The resulting eight treatment combinations are shown in Table 9.4 (left). We see that in addition to the two generators, we also have a further highly undesirable confounding of the main effect of **A** with the grand mean: the column A only contains the high level. This is a consequence of the interplay of the two generators, and we find this additional confounding directly by comparing the left- and right-hand side of their generalized interaction:

$$G_1 G_2 = ABCDE \cdot BCDE = ABBCCDDEE = A = +1 .$$

Some further trial-and-error reveals that no useful second generator is available if we confound the five-way interaction with the grand mean in our first generator. A reasonably good pair of generators uses two three-way interactions, such as

$$G_1 : ABD = +1 \quad \text{and} \quad G_2 : ACE = +1 ,$$

with generalized interaction

$$G_1 G_2 = AABCDE = BCDE = +1 .$$

The resulting treatment combinations are shown in Table 9.4 (right). We note that some—but not all—main effects and two-way interactions are now confounded.

Finding good pairs of generators is not entirely straightforward, and software or tabulated designs are often used.

9.5.2   A Real-Life Example—Yeast Medium Composition

Recall that we used a 2^{5-1} half-replicate for our yeast medium example in Sect. 9.4, but that we had to remove all observations with **N2** at the low level from the first replicate of this experiment. This effectively introduces a second generator for this replicate, namely $C = +1$. Since **N2** is only observed on one level, no effects involving this factor can be estimated. In addition, the combination of the second generator with the original generator $ABCDE = +1$ leads to the additional alias $AB = DE$ between the interaction **Glc:N1** and the interaction **Vit1:Vit2** for this replicate. Fortunately, the corresponding observations from the second replicate were not affected by this problem, such that the pooled data from both replicates could be analyzed as planned.

9.5.3   A Real-Life Example—2^{7-2}-Fractional Factorial

The transformation of yeast cells is an important experimental technique, but many protocols have very low yield. In an attempt to define a more reliable and efficient protocol, seven treatment factors were considered in combination: Ion, PEG, DMSO, Glycerol, Buffer, EDTA, and amount of carrier DNA. With each component in two concentrations, the full treatment structure is a 2^7-factorial with 128 treatment combinations. This experiment size is prohibitive since each treatment requires laborious subsequent steps, but 32 treatment combinations were considered reasonable for implementing this experiment. This requires a quarter-replicate of the full design.

Ideally, we want to find two generators that alias main effects and two-way interactions with interactions of order three and higher, but no such pair of generators exists in this case. We are confronted with the problem of confounding *some* two-way interactions with each other, while other two-way interactions are confounded with three-way interactions.

Preliminary experiments suggested that the largest interactions involve Ion, PEG, and potentially Glycerol, while interactions with other components are small. A reasonable design uses the two generators

$$G_1 : ABCDF = +1 \text{ and } G_2 : ABDEG = +1$$

with generalized interaction $CF = EG$.

The two-factor interactions involving the factors **C**, **E**, **F**, and **G** are then confounded with each other, while two-way interactions involving the remaining factors **A**, **B**, and **D** are confounded with interactions of order three or higher. Hence, selecting **A**, **B**, **D** as the factors Ion, PEG, and Glycerol allows us to create a design with 32 treatment combinations that reflects our subject-matter knowledge and allows estimation of all relevant two-way interactions while confounding those two-way interactions that we consider negligible. For example, we cannot disentangle an interaction of DMSO and EDTA from an interaction of Buffer and carrier DNA, but this does not jeopardize the interpretation of this experiment.

9.6 Characterizing Fractional Factorials

Two measures to characterize the severity of confounding in a fractional factorial design are the *resolution* and the *aberration*.

9.6.1 Resolution

A fractional factorial design has *resolution K* if the grand mean is confounded with at least one factor of order K, and no factor of lower order. The order is typically given as a Roman numeral. For example, a 2^{3-1} design with generator $ABC = +1$ has order III, and we denote such a design as 2^{3-1}_{III}.

For a factor of any order, the resolution gives the lowest order of a factor confounded with it: a resolution-III design confounds main effects with two-way interactions (III $= 1 + 2$), and the grand mean with a three-way interaction (III $= 0 + 3$). A resolution-V design confounds main effects with four-way interactions (V $= 1 + 4$), two-way interactions with three-way interactions (V $= 2 + 3$), and the five-way interaction with the grand mean (V $= 5 + 0$).

Designs with more factors allow fractions of higher resolution. Our previous 2^5-factorial example admits a 2^{5-1}_{V} design with 16 combinations, and a 2^{5-2}_{III} design with 8 combinations. With the first design, we can estimate main effects and two-way interactions free of other main effects and two-way interactions, while the second design aliases main effects with two-way interactions. Our 7-factor example has resolution IV.

In practice, resolutions III, IV, and V are the most ubiquitous, and a resolution of V is often the most useful if it is achievable, since then main effects and two-way interactions are aliased only with interactions of order three and higher. Main effects and two-way interactions are confounded for resolution III, and these designs are useful for screening larger numbers of factors, but usually not for experiments where relevant information is expected in the two-way interactions. If a design has many treatment factors, we can also construct fractions with resolution higher than V, but it might be more practical to further reduce the experiment size and use an additional generator to construct a design with resolution V, for example.

Resolution IV confounds two-way interactions with each other. While this is rarely desirable, we might find multiple generators that leave some two-way interactions unconfounded with other two-way interactions, as in our 7-factor example. Such designs offer dramatic decreases in the experiment size for large numbers of factors. For example, full factorials for nine, 10, and eleven factors have 512, 1024, and 2048 treatment combinations, respectively. For most experiments, this is not practically implementable. However, fractional factorials of resolution IV only require 32 treatment combinations in each case, which is a very attractive proposition in many situations.

Similarly, a 2^{7-2} design has resolution IV, since some of the two-way interactions are confounded. The maximal resolutions for the 2^7 series are $2_{VII}^{7-1}, 2_{IV}^{7-2}, 2_{IV}^{7-3}, 2_{III}^{7-4}$. Thus, the resolution drops with increasing fraction, and not all resolutions might be achievable for a given number of factors.

9.6.2 Aberration

For the 2^7-factorial, both one-quarter and one-eighth reductions lead to a resolution-IV design, even though these designs have very different severity of confounding.

The *aberration* provides an additional criterion to compare designs with identical resolution. It is based on the idea that we prefer aliasing higher-order interactions to aliasing lower-order interactions.

We find the aberration of a design as follows: we write down the generators and derive their generalized interactions. We then sort the resulting set of alias relations by word length and count how many relations there are of each length. The fewer words of short length a set of generators produces, the more we would prefer it over a set with more short words.

For example, the two generators

$$ABCDE = +1 \quad \text{and} \quad ABCEG = +1$$

yield a 2_{IV}^{7-2} design with generalized interaction $ABCDE \cdot ABCEG = DG = +1$. This design has a set of generating alias relations with one word of length two, and two words of length five.

The two generators

$$ABCF = +1 \quad \text{and} \quad ADEG = +1 .$$

also yield a 2_{IV}^{7-2} design, this time with generalized interaction $ABCF \cdot ADEG = BCDEFG = +1$. The corresponding aliases thus contain two words of length four and one word of length six and we would prefer this set of generators over the first set because of its less severe confounding.

9.7 Factor Screening

A common problem, especially at the beginning of designing an assay or investigating any system, is to determine which of the vast number of possible factors actually have a relevant influence on the response. For example, we might want to design a toxicity assay with a luminescent readout on a 48-well plate, where luminescence is supposed to be directly related to the number of living cells in each well, and is thus a proxy for toxicity of a substance pipetted into a well. Apart from the substance's concentration and toxicity, there are many other factors that we might imagine can influence the readout. Examples include the technician, amount of shaking before reading, the reader type, batch effects of chemicals, temperature, setting time, labware, type of pipette (small/large volume), and many others.

Before designing an experiment for more detailed analyses of relevant factors, we may want to conduct a *factor screening* to determine which factors are *active* and appreciably affect the response. Subsequent experimentation then only includes the active factors and, having reduced the number of treatment factors, can be designed with the methods previously discussed.

Factor screening designs make extensive use of the assumption that the proportion of active factors among those considered is small. We usually also assume that we are only interested in the main effects and can ignore the interaction effects for the screening. This assumption is justified because we will not make any inference on *how exactly* the factors influence the response, but are for the moment only interested in discarding factors of no further interest.

9.7.1 Fractional Factorials

One class of screening designs uses fractional factorials of resolution III. Noteworthy examples are the 2_{III}^{15-11} design, which allows screening 15 factors in 16 runs, or the 2_{III}^{31-26} design, which allows screening 31 factors in 32 runs!

A problem of this class of designs is that the 'gap' between useful screening designs increases with increasing number of factors, because we can only consider fractions that are powers of two: reducing a 2^7-design with 128 runs yields designs of 64 runs (2^{7-1}) and 32 runs (2^{7-2}), but we cannot find designs with less than 64 and more than 32 runs, for example. On the other hand, fractional factorials are familiar designs that are relatively easy to interpret and if a reasonable design is available, there is no reason not to consider it.

Factor screening experiments will typically use a single replicate of a (fractional) factorial, and effects cannot be tested formally. If only a minority of factors is active, we can use the method by Lenth to still identify the active factors by more informal comparisons (Lenth 1989); see Sect. 6.4.2 for details on this method.

9.7.2   Plackett–Burman Designs

A different idea for constructing screening designs was proposed by Plackett and Burman in a seminal paper (Plackett and Burman 1946). These designs require that the number of runs is a multiple of four. The most commonly used are the designs in 12, 20, 24, and 28 runs, which can screen 11, 19, 23, and 27 factors, respectively. Plackett–Burman designs do *not* have a simple confounding structure that could be determined with generators. Rather, they are based on the idea of partially confounding some fraction of each effect with other effects. These designs are used for screening main effects only, as main effects are already confounded with two-way interactions in rather complicated ways that cannot be easily disentangled by follow-up experiments. Plackett–Burman designs considerably increase the available options for the screening experiment sizes, and offer designs when no fractional factorial design is available.

9.8   Blocking Factorial Experiments

With many treatments, blocking a design becomes challenging because the efficiency of blocking deteriorates with increasing block size, and there are other limits on the maximal number of units per block. The incomplete block designs in Sect. 7.3 are a remedy for this problem for unstructured treatment levels. The idea of fractional factorial designs is useful for blocking factorial treatment structures and exploits their properties by deliberately confounding (higher-order) interactions with block effects. This reduces the required block size to the size of the corresponding fractional factorial.

We can further extend this idea by using different confoundings for different sets of blocks, such that each set accommodates a different fraction of the same factorial treatment structure. We are then able to recover most of the effects of the full factorial, albeit with different precision.

We consider a blocked design with a 2^3-factorial treatment structure in blocks of size four as our main example. This is a realistic scenario if studying combinations of three treatments on mice and blocking by litter, with typical litter sizes being below eight. Two questions arise: (i) which treatment combinations should we assign to the same block? and (ii) with replication of blocks, should we use the same assignment of treatment combinations to blocks? If not, how should we determine treatment combinations for sets of blocks?

9.8.1 Half-Fraction

A first idea is to use a half-replicate of the 2^3-factorial and assign its four treatment combinations to the four units in each block. For example, we can use the generator $ABC = +1$ and randomize the same treatment combinations $\{a, b, c, abc\}$ independently within each block. A layout for four blocks is

Block	Generator	1	2	3	4
I	$ABC = +1$	a	b	c	abc
II	$ABC = +1$	a	b	c	abc
III	$ABC = +1$	a	b	c	abc
IV	$ABC = +1$	a	b	c	abc

This design confounds the three-way interaction with the block effect and resembles a replication of the same fractional factorial, where systematic differences between replicates are accounted for by the block effects. The fractional factorial has resolution III, and main effects are confounded with two-way interactions within each block (and thereby also overall).

From the 16 observations, we require four degrees of freedom for estimating the treatment parameters, and three degrees of freedom for the block effect, leaving us with nine residual degrees of freedom. The latter can be increased by using more blocks, where we gain four observations with each block and lose one degree of freedom for the block effect. Since the effect aliases are the same in each block, increasing the number of blocks does not change the confounding: no matter how many blocks we use, we are unable to disentangle the main effect of **A**, say, and the **B:C** interaction.

9.8.2 Half-Fraction with Alternating Replication

We can improve the design substantially by noting that it is not required to use the same half-replicate in each block. For instance, we might instead use the generator $ABC = +1$ with combinations $\{a, b, c, abc\}$ to create a half-replicate of the treatment structure for the first two of four blocks, and use the corresponding generator $ABC = -1$ (the *fold-over*) with combinations $\{(1), ab, ac, bc\}$ for the remaining two blocks.

With two replicates for each of the two levels of the three-way interaction, its parameters are estimable using the block totals. All other effects can be estimated more precisely, since we now have two replicates of the full factorial design after we account for the block effects.

The corresponding assignment is

Block	Generator	1	2	3	4
I	$ABC = +1$	a	b	c	abc
II	$ABC = +1$	a	b	c	abc
III	$ABC = -1$	(1)	ab	ac	bc
IV	$ABC = -1$	(1)	ab	ac	bc

and shows that while the half-fraction of a 2^3-factorial is not an interesting option in itself due to the severe confounding, it gives a very appealing design for reducing block sizes.

For example, we have confounding of **A** with **B:C** for observations based on the $ABC = +1$ half-replicates (with $A = BC$), but we can resolve this confounding using observations from the other half-replicate, for which $A = -BC$. Indeed, for blocks I and II, the estimate of the **A** main effect is $(a + abc) - (b + c)$ and for blocks III and IV it is $(ab + ac) - (bc + (1))$. Similarly, the estimates for **B:C** are $(a + abc) - (b + c)$ and $(bc + (1)) - (ab + ac)$, respectively. Note that these estimates are all free of block effects. Then, the estimates of the two effects are also free of block effects and are proportional to $[(a + abc) - (b + c)] + [(ab + ac) - (bc + (1))] = (a + ab + ac + abc) - ((1) + b + c + bc)$ for **A**, respectively, $[(a + abc) - (b + c)] - [(ab + ac) - (bc + (1))] = ((1) + a + bc + abc) - (b + c + ab + ac)$ for **B:C**. These are the same estimates as for a two-fold replicate of the full factorial design. Somewhat simplified: the first two blocks allow estimation of the sum of **A** main effect and **B:C** interaction, while the second pair allows estimation of their difference. The sum of these two estimates is $2 \cdot A$, while the difference is $2 \cdot BC$.

The same argument does not hold for the **A:B:C** interaction, of course. Here, we have to contrast observations in $ABC = +1$ blocks with observations in $ABC = -1$ blocks, and block effects do not cancel. If instead of four blocks, our design only uses two blocks—one for each generator—then main effects and two-way interactions can still be estimated, but the three-way interaction is completely confounded with the block effect.

Using a classical ANOVA for the analysis, we find two error strata for the inter- and intra-block errors, and the corresponding F-test for **A:B:C** in the inter-block stratum with two denominator degrees of freedom: we have four blocks, and lose one degree of freedom for the grand mean, and one degree of freedom for the **A:B:C** parameters. All other tests are in the intra-block stratum and based on six degrees of freedom: a total of $4 \cdot 4 = 16$ observations, with seven degrees of freedom spent on the model parameters except the three-way interaction, and three degrees of freedom spent on the block effects.

A useful consequence of these considerations is the possibility of augmenting a fractional factorial design with the complementary half-replicate. For example, we might consider a half-replicate of a 2^5-factorial with generator $ABCDE = +1$. If we find large effects for the confounded two- and three-way interactions, we can use a single second experiment with $ABCDE = -1$ to provide the remaining treatment combinations and disentangle these interactions. We account for systematic differences between the two experiments by introducing a block with two levels, confounded with the five-way interaction.

9.8.3 Excursion: Split-Unit Designs

While using the highest-order interaction to define the confounding with blocks is the natural choice, we could also use any other generator. In particular, we might use $A = +1$ and $A = -1$ as our two generators, thereby allocating half the blocks to the low level of **A**, and the other half to its high level. In other words, we randomize **A** on the block factor, and the remaining treatment factors are randomized within each block. This is precisely the split-unit design with the blocking factor as the whole-unit factor, and **A** randomized on it. With four blocks, we need one degree of freedom to estimate the block effect, and the remaining three degrees of freedom are split into estimating the **A** main effect (1 d.f.) and the between-block residual variance (2 d.f.). All other treatment effects profit from the removal of the block effect and are tested with 6 degrees of freedom for the within-block residual variance.

The use of generators offers more flexibility than a split-unit design, because it allows us to confound *any* effect with the blocking factor, not just a main effect. Whether this is an advantage depends on the experiment: if application of the treatment factors to experimental units is equally simple for all factors, then it is usually more helpful to confound a higher-order interaction with the blocking factor. This design then allows estimation of all main effects and their contrasts with equal precision, and lower-order interaction effects can also be estimated precisely. A split-unit design, however, offers advantages for the logistics of the experiment if levels of one treatment factor are more difficult to change than levels of the other factors. By confounding the hard-to-change factor with the blocking factor, the experiment becomes easier to implement. Split-unit designs are also conceptually simpler than confounding of interaction effects with blocks, but that should not be the sole motivation for using them.

9.8.4 Half-Fraction with Multiple Generators

We are often interested in all effects of a factorial treatment design, especially if this design has only few factors. Using a single generator and its fold-over, however, provides much lower precision for the corresponding effect, which might be undesirable. An alternative strategy is to use *partial confounding* of effects with blocks by employing *different* generators and their fold-overs for different pairs of blocks.

For example, we consider again the half-replicate of a 2^3-factorial, with four units per block. If we have resources for 32 units in eight blocks, we can form four pairs of blocks and confound a different effect in each pair by using the generators $G_1 : ABC = \pm 1$ for our first, $G_2 : AB = \pm 1$ for the second, $G_3 : AC = \pm 1$ for the third, and $G_4 : BC = \pm 1$ for the fourth pair of blocks:

Block	Generator	1	2	3	4
I	$ABC = +1$	a	b	c	abc
II	$ABC = -1$	(1)	ab	ac	bc
III	$AB = +1$	(1)	c	ab	abc
IV	$AB = -1$	a	b	ac	bc
V	$AC = +1$	(1)	b	ac	abc
VI	$AC = -1$	a	b	ab	bc
VII	$BC = +1$	(1)	a	bc	abc
VIII	$BC = -1$	b	c	ab	ac

Information about each interaction is now contained in the inter-block error stratum and the residual (intra-block) error stratum of the ANOVA:

	Df	Sum Sq	Mean Sq	F value	Pr(>F)
Error stratum: Block					
A:B	1	0.04	0.04	0.28	6.32e-01
A:C	1	2.56	2.56	18.46	2.32e-02
B:C	1	2.04	2.04	14.75	3.11e-02
A:B:C	1	1.95	1.95	14.09	3.30e-02
Residuals	3	0.42	0.14		
Error stratum: Within					
A	1	1.79	1.79	1.8	1.97e-01
B	1	0	0	0	9.55e-01
C	1	1.4	1.4	1.41	2.52e-01
A:B	1	3.67	3.67	3.69	7.17e-02
A:C	1	1.62	1.62	1.63	2.19e-01
B:C	1	0.07	0.07	0.07	7.97e-01
A:B:C	1	0.01	0.01	0.01	9.23e-01
Residuals	17	16.91	0.99		

In this design, each two-way interaction can be estimated using within-block information of three pairs of blocks, and the same is true for the three-way interaction. Additional estimates can be defined based on the inter-block information, similar to a BIBD. The inter- and intra-block estimates can be combined, but this is rarely done in practice for a classic ANOVA, where the more precise within-block estimates are often used exclusively. In contrast, linear mixed models offer a direct way of basing all estimates on all available data; a corresponding model for this example is specified as y~A*B*C+(1|block).

9.8.5 Multiple Aliasing

We can further reduce the required block size by considering higher fractions of a factorial. As we saw in Sect. 9.5, these require several simultaneous generators, and additional aliasing occurs due to the generalized interaction between the generators.

For example, the half-fraction of a 2^5-factorial still requires a block size of 16, which might not be practical. We further reduce the block size using the two pairs of generators

$$ABC = \pm 1, \quad ADE = \pm 1,$$

with generalized interaction $ABC \cdot ADE = BCDE$, leading to a 2^{5-2} treatment design (Finney 1955, p102). Each of the four combinations of these two pairs selects eight of the 32 possible treatment combinations and a single replicate of this design requires four blocks:

Block	Generator		1	2	3	4	5	6	7	8
I	$ABC = -1, ADE = -1$	(1)	bc	de	bcde	abd	acd	abe	ace
II	$ABC = -1, ADE = +1$	b	c	bde	cde	ad	abcd	ae	abce
III	$ABC = +1, ADE = -1$	d	bcd	e	bce	ab	ac	abde	acde
IV	$ABC = +1, ADE = +1$	bd	cd	be	ce	a	abc	ade	abcde

In this design, the two three-way interactions **A:B:C** and **A:D:E** and their generalized four-way interaction **B:C:D:E** are partially confounded with block effects. All other effects, and in particular all main effects and all two-way interactions, are free of block effects and estimated precisely. By carefully selecting the generators, we are often able to confound effects that are known to be of limited interest to the researcher.

9.8.6 A Real-Life Example—Proteomics

As a concrete example of blocking a factorial design, we discuss a simplified variant of a proteomics study in mice. The main target of this study is the response to inflammation, and a drug is available to trigger this response. One pathway involved in the response is known and many of the proteins involved as well as the receptor upstream of the pathway have been identified. However, related experiments suggested that the drug also activates alternative pathways involving other receptors, and one goal of the experiment is to identify proteins involved in these pathways.

The experiment has three factors in a 2^3-factorial treatment design: administration of the drug or a placebo, a short or long waiting time between drug administration and measurements, and the use of the wild-type or a mutant receptor for the known pathway, where the mutant inhibits binding of the drug and hence deactivates the pathway.

Expected Results

We can broadly distinguish three classes of proteins that we expect to find in this experiment.

The first class is proteins *directly involved* in the known pathway. For these, we expect low levels of abundance for a placebo treatment, because the placebo does not activate the pathway. For the drug treatment, we expect to see high abundance in the wild-type, as the pathway is then activated, but low abundance in the mutant, since the drug cannot bind to the receptor and thus pathway activation is impeded. In other words, we expect a large genotype-by-drug interaction.

The second class is proteins in the *alternative* pathway(s) activated by the drug but exhibiting a different receptor. Here, we would expect to see high abundance in both wild-type and mutant for the drug treatment and low abundance in both genotypes for a placebo treatment, since the mutation does not affect receptors in these pathways. This translates into a large drug main effect, but no genotype main effect and no genotype-by-drug interaction.

The third class is proteins *unrelated* to any mechanisms activated by the drug. Here, we expect to see the same abundance levels in both genotypes for both drug treatments, and no treatment factor should show a large and significant effect.

We are somewhat unsure what to expect for the duration. It seems plausible that a protein in an activated pathway will show lower abundance after longer time, since the pathway should trigger a response to the inflammation and lower the inflammation. This would mean that a three-way interaction exists at least for proteins involved in the known or alternative pathways. A different scenario results if one pathway takes longer to activate than another pathway, which would present as a two- or three-way interaction of drug and/or genotype with the duration.

Mass Spectrometry Using Tags

Absolute quantification of protein abundances is very difficult to achieve. An alternative technique is to use mass spectrometry with *tags*, small molecules that attach to each protein and modify its mass by a known amount. With four different tags available, we can then pool all proteins from four different experimental conditions and determine their *relative* abundances by comparing the four resulting peaks in the mass spectrum for each protein.

We have 16 mice available, eight wild-type and eight mutant mice. Since we have eight treatment combinations but only four tags, we need to block the experiment in sets of four. An obvious candidate is confounding the block effect with the three-way interaction genotype-by-drug-by-time. This choice is shown in Fig. 9.3, and each label corresponds to a treatment combination in the first two blocks and the opposite treatment combination in the remaining two blocks.

The main disadvantage of this choice is the confounding of the three-way interaction with the block effect, which only allows imprecise estimation, and it is unlikely that the effect sizes are large enough to allow reliable detection in this design. Alternatively, we can use two generators for the two pairs of blocks, the first confounding

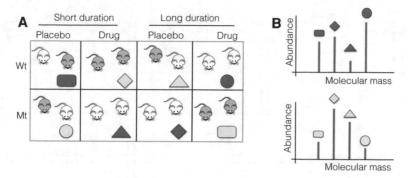

Fig. 9.3 Proteomics experiment. **A** 2^3-factorial treatment structure with three-way interaction confounded in two blocks. **B** Mass spectra with four tags (symbol) for same protein from two blocks (shading)

the three-way interaction, and the second confounding one of the three two-way interactions. A promising candidate is the drug-by-duration interaction, since we are very interested in the genotype-by-drug interaction and would like to detect different activation times between the known and alternative pathways, but we do not expect a drug-by-duration interaction of interest. This yields the data shown in Fig. 9.4, where the eight resulting protein abundances are shown separately for short and long duration between drug administration and measurement, and for three typical proteins in the known pathway, in an alternative pathway, and unrelated to the inflammation response.

9.9 Notes and Summary

Notes

Deliberate effect confounding in factorial designs was fully developed in the 1940s (Fisher 1941; Finney 1945) and is an active research area to this day. A general review is given in Gunst and Mason (2009), and modern developments for multistratum designs are given in Cheng (2019). Some specific designs are discussed for engineering applications in Box (1992) and Box and Bisgaard (1993).

Fractional factorials can also be constructed for factors with more than two levels, such as the 3^k-series (Cochran and Cox 1957), or generally the p^k-series (p a prime number). A more general concept for confounding in factorials with mixed number of factor levels are *design keys* (Patterson and Bailey 1978). For the analysis of non-replicated designs, the methods by Lenth (1989) (discussed in Sect. 6.4.2) and Box and Meyer (1986) are widely used.

The website for NIST's *Engineering Statistics Handbook* provides tables with commonly used 2^{k-l}-fractional factorials, Plackett–Burman, and other useful designs in its Chap. 5.

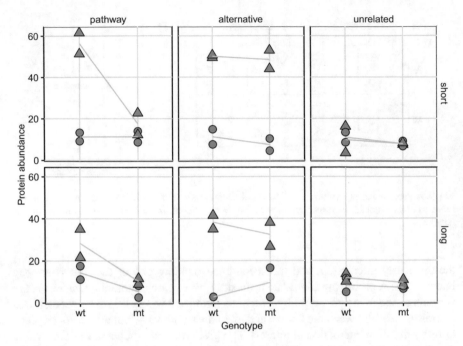

Fig. 9.4 Data of proteomics experiment. Round point: placebo, triangle: drug treatment. Panels show typical protein scenarios in columns and waiting duration in rows

Aliasing and the Linear Model

We provide more details on the aliasing in a half-fraction of the 2^3-factorial design with generic treatment factors **A**, **B**, and **C**, each with levels -1 (low) and $+1$ (high). The linear model for this design is

$$
\begin{aligned}
y_{ijkl} =\mu &+ \alpha_A \cdot a_i + \alpha_B \cdot b_j + \alpha_C \cdot c_k \\
&+ \alpha_{AB} \cdot a_i \cdot b_j + \alpha_{AC} \cdot a_i \cdot c_k + \alpha_{BC} \cdot b_j \cdot c_k \\
&+ \alpha_{ABC} \cdot a_i \cdot b_j \cdot c_k + e_{ijkl} \,,
\end{aligned}
$$

where a_i, b_j, c_k encode the factor level of **A**, **B**, and **C**, respectively, for that specific observation. With a sum-encoding, we have $a_i = -1$ if **A** is on the low level, and $a_i = +1$ if **A** is on the high level, with values for b_j, c_k accordingly. The seven parameters α_X are the effects of the corresponding factor that we want to estimate.

Using the generator $ABC = +1$ then translates to imposing the relation $a_i \cdot b_j \cdot c_k = +1$ for each observation i, j, k, and we can replace $a_i \cdot b_j \cdot c_k$ with $+1$ in the linear model equation. It follows that the parameter α_{ABC} of the three-way interaction is completely confounded with the grand mean μ. Similarly, we note that $a_i \cdot b_j = c_k$ for each observation, and we can replace $a_i \cdot b_j$ with c_k in the model equation. Thus, the two parameters α_{AB} and α_C, encoding the effect of the two-way interaction **A:B** and the main effect of **C**, respectively, are completely confounded and only their sum $\alpha_{AB} + \alpha_C$ can be estimated. Continuing this way, we find that the generator implies the linear model

$$y_{ijkl} = \beta_0 + \beta_1 \cdot a_i + \beta_2 \cdot b_j + \beta_3 \cdot c_k + e_{ijkl} \, ,$$

and we can only estimate its four *derived parameters* $\beta_0 = \mu + \alpha_{ABC}$, $\beta_1 = \alpha_A + \alpha_{BC}$, $\beta_2 = \alpha_B + \alpha_{AC}$, and $\beta_3 = \alpha_C + \alpha_{AB}$, each parameter corresponding to one alias set.

Similarly, the generator $ABC = -1$ implies that $a_i \cdot b_j = -c_k$, for example, leading to the four derived parameters $\gamma_0 = \mu - \alpha_{ABC}$, $\gamma_1 = \alpha_A - \alpha_{BC}$, $\gamma_2 = \alpha_B - \alpha_{AC}$, and $\gamma_3 = \alpha_C - \alpha_{AB}$ as the estimable quantities for this half-fraction of the 2^3-factorial design.

Using R

The two R packages FrF2 (Grömping 2014) and PLANOR (Kobilisnsky et al. 2012; Kobilinsky et al. 2017) provide functionality to construct and analyze fractional factorial designs. A simple semi-manual way is to generate the 2^k rows of the full factorial with entries ± 1, adding a column for the generator defined by multiplication of the corresponding columns, and removing all rows where this column is -1, for example. The FrF2 package also provides the pb() method for constructing Plackett–Burman designs. The *Comprehensive R Archive Network (CRAN)* maintains a list of DoE-related R packages.

Summary

Fractional factorial designs reduce the experiment size when using many treatment factors. In a 2^k-factorial, all k treatment factors have two levels; a formal generator algebra can then be used to define fractional replicates and provides the alias sets of confounded parameters. The resolution measures the degree of confounding. Higher fractions require more than one generator, and greater care is required to arrive at a useful confounding in this case.

Factor screening aims at identifying the active factors from a (potentially large) set of treatment factors, such that subsequent experiments can focus on relevant treatment factors. Fractional factorials are commonly used for this task, and designs with low resolution are often adequate. The Plackett–Burman designs fill gaps in available experiment sizes.

Fractional factorials also offer advantages when the block size does not accommodate the full treatment structure. Effects can then be partially deconfounded by using different generators for pairs of blocks.

References

Box, G. E. P. (1992). "What can you find out from sixteen experimental runs?" In: Quality Engineering 5.1, pp. 167–178.

Box, G. E. P. and R. D. Meyer (1986). "An analysis for unreplicated fractional factorials". In: Technometrics 28.1, pp. 11–18.

Box, G. E. P. and S. Bisgaard (1993). "Quality quandaries: iterative analysis from two-level factorials". In: Quality Engineering 6.2, pp. 319–330.

Cheng, C.-S. (2019). Theory of Factorial Design: Single- and Multi-Stratum Experiments. Chapman & Hall/CRC.

Cochran, W. G. and G. M. Cox (1957). Experimental Designs. John Wiley & Sons, Inc.

Finney, D. J. (1945). "The fractional replication of factorial arrangements". In: Annals of Eugenics 12, pp. 291-301.

Finney, D. J. (1955). Experimental Design and its Statistical Basis. The University of Chicago Press.

Fisher, R. A. (1941). "The theory of confounding in factorial experiments in relation to the theory of groups". In: Annals of Human Genetics 11.1, pp. 341-353.

Grömping, U. (2014). "R package FrF2 for creating and analyzing fractional factorial 2-level designs". In: Journal of Statistical Software 56.1, e1–e56.

Gunst, R. F. and R. L. Mason (2009). "Fractional factorial design". In: WIREs Computational Statistics 1, pp. 234–244.

Kobilisnsky, A., A. Bouvier, and H. Monod (2012). PLANOR: an R package for the automatic generation of regular fractional factorial designs. Tech. rep. INRA, e1–e97.

Kobilinsky, A., H. Monod, and R. A. Bailey (2017). "Automatic generation of generalised regular designs." In: Computational Statistics and Data Analysis 113, pp. 311–329.

Lenth, R. V. (1989). "Quick and easy analysis of unreplicated factorials". In: Technometrics 31.4, pp. 469–473.

Patterson, H. D. and R. A. Bailey (1978). "Design keys for factorial experiments". In: Journal of the Royal Statistical Society C 27.3, pp. 335–343.

Plackett, R. L. and J. P. Burman (1946). "The design of optimum multifactorial experiments". In: Biometrika 33.4, pp. 305–325.

Chapter 10
Experimental Optimization
with Response Surface Methods

10.1 Introduction

In previous chapters, we were mostly concerned with comparative experiments using *categorical* or *qualitative* factors, where each factor level is a discrete category and there is no specific order in the factor levels. We took a small detour into the analysis of ordered factors in Sect. 5.2.6. In this chapter, we consider designs to explore and quantify the effects of several *quantitative* factors on a response, where factor levels are chosen from a potentially infinite range of possible levels. Examples of such factors are concentrations, temperatures, and durations, whose levels can be chosen arbitrarily from a continuum.

We are interested in two main questions: (i) to describe how smoothly changing the level of one or more factors affects the response and (ii) to determine factor levels that maximize the response. For addressing the first question, we consider designs that experimentally determine the response at specific points around a given experimental condition and use a regression model to interpolate or extrapolate the expected response values at other points. For addressing the second question, we first explore how the response changes around a given setting, determine the factor combinations that likely yield the highest increase in the response, experimentally evaluate these combinations, and then start over by exploring around the new 'best' setting. This idea of *sequential experimentation* for optimizing a response is often implemented using *response surface methodologies (RSM)*. An example application is finding optimal conditions for temperatures, pH, and flow rates in a bioreactor to maximize yield of a process.

Here, we revisit our yeast medium example from Sect. 9.4, and our goal is to find the concentrations (or amounts) of five ingredients—glucose **Glc**, two nitrogen sources **N1** and **N2**, and two vitamin sources **Vit1** and **Vit2**—that maximize the growth of a yeast culture.

© Springer Nature Switzerland AG 2021
H.-M. Kaltenbach, *Statistical Design and Analysis of Biological Experiments*,
Statistics for Biology and Health, https://doi.org/10.1007/978-3-030-69641-2_10

10.2   Response Surface Methodology

The key new idea is to consider the response as a smooth function of the quantitative treatment factors. We generically call the treatment factor levels $x_1, \ldots, x_k$ for k factors, so that we have five such variables for our example, corresponding to the five concentrations of medium components. We again denote the response variable by y, which is the increase in optical density in a defined time-frame in our example. The *response surface* $\phi(\cdot)$ relates the expected response to the experimental variables:

$$\mathbb{E}(y) = \phi(x_1, \ldots, x_k) \ .$$

We assume that small changes in the variables will yield small changes in the response, so the surface described by $\phi(\cdot)$ is smooth enough that, given two points and their resulting responses, we feel comfortable in interpolating intermediate responses. The shape of the response surface and its functional form $\phi(\cdot)$ are unknown, and each measurement of a response is additionally subject to some variability.

The goal of a *response surface design* is to define n design points $(x_{1,j}, \ldots, x_{k,j})$, $j = 1 \ldots n$, and use a reasonably simple yet flexible regression function $f(\cdot)$ to approximate the true response surface $\phi(\cdot)$ from the resulting measurements y_j so that $f(x_1, \ldots, x_k) \approx \phi(x_1, \ldots, x_k)$, at least locally around a specified point. This approximation allows us to predict the expected response for any combination of factor levels that is not too far outside the region that we explored experimentally.

Having found such approximation, we can determine the *path of steepest ascent*, the direction along which the response surface increases the fastest. For optimizing the response, we design the next experiment with design points along this gradient (the *gradient pursuit* experiment). Having found a new set of conditions that gives higher responses than our starting condition, we iterate the two steps: locally approximate the response surface around the new best treatment combination and follow its path of steepest ascent in a subsequent experiment. We repeat these steps until no further improvement of the measured response is observed, we found a treatment combination that yields a satisfactorily high response, or we run out of resources. This idea of a *sequential* experimental strategy is illustrated in Fig. 10.1 for two treatment factors whose levels are on the horizontal and vertical axis, and a response surface shown by its contours.

For implementing this strategy, we need to decide what point to start from; how to approximate the surface $\phi(\cdot)$ locally; how many and what design points $(x_{1,j}, \ldots, x_{k,j})$ to choose to estimate this approximation; and how to determine the path of steepest ascent.

Optimizing treatment combinations usually means that we already have a reasonably reliable experimental system, so we use the current experimental condition as our initial point for the optimization. An important aspect here is the reproducibility of the results: if the system or the process under study does not give reproducible results at the starting point, there is little hope that the response surface experiments will give improvements.

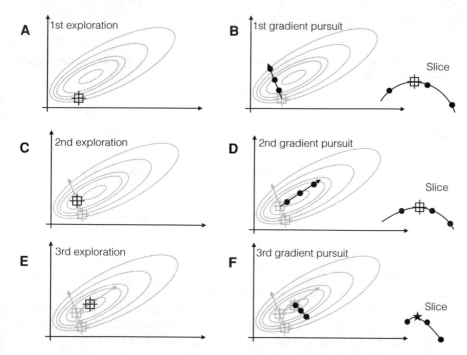

Fig. 10.1 Sequential experiments to determine optimum conditions. Three exploration experiments (**A, C, E**), each followed by a gradient pursuit (**B, D, F**). Dotted lines: contours of response surface. Black lines and dots: region and points for exploring and gradient pursuit. Inlet curves in panels **B, D, F** show the slice along the gradient with measured points and resulting maximum

The remaining questions are interdependent: we typically choose a simple (polynomial) function to locally approximate the response surface. Our experimental design must then allow the estimation of all parameters of this function, the estimation of the residual variance, and ideally provide information to separate residual variation from lack of fit of the approximation. Determining a gradient is straightforward for a polynomial interpolation function.

We discuss two commonly used models for approximating the response surface: the *first-order model* requires only a simple experimental design but does not account for curvature of the surface and predicts ever-increasing responses along the path of steepest ascent. The *second-order model* allows for curvature and has a defined stationary point (a maximum, minimum, or saddle-point), but requires a more complex design for estimating its parameters.

10.3 The First-Order Model

We start our discussion by looking into the first-order model for locally approximating the response surface.

10.3.1 Model

The first-order model for k quantitative factors (without interactions) is

$$y = f(x_1, \ldots, x_k) + e = \beta_0 + \beta_1 x_1 + \cdots + \beta_k x_k + e = \beta_0 + \sum_{i=1}^{k} \beta_i x_i + e \ .$$

We estimate its parameters using standard linear regression; the parameter β_i gives the amount by which the expected response increases if we increase the ith factor from x_i to $x_i + 1$, keeping all other factors fixed. Without interactions, the predicted change in response is independent of the values of all other factors, and interactions could be added if necessary. We assume a constant error variance $\text{Var}(e) = \sigma^2$ and justify this by the fact that we explore the response surface only locally.

10.3.2 Path of Steepest Ascent

At any given point $(x_1^*, \ldots, x_k^*)$, the gradient of $f(\cdot)$ is

$$g(x_1^*, \ldots, x_k^*) = (\beta_1, \ldots, \beta_k)^T \ ,$$

and gives the direction with fastest increase in response. This gradient is independent of the factor levels (the function $g(\cdot)$ does not depend on any x_i), and the direction of steepest ascent is the same no matter where we start.

In the next iteration, we explore this direction experimentally to find a treatment combination that yields a higher expected response than our starting condition. The local approximation of the true response surface by our first-order model is likely to fail once we venture too far from the starting condition, and we will encounter decreasing responses and increasing discrepancies between the response predicted by the approximation and the response actually measured. One iteration of exploration and gradient pursuit is illustrated in Fig. 10.2.

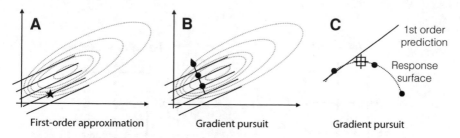

Fig. 10.2 Sequential response surface experiment with first-order model. **A** The first-order model (solid lines) build around a starting condition (star) approximates the true response surface (dotted lines) only locally. **B** The gradient follows the increase of the plane and predicts indefinite increase in response. **C** The first-order approximation (solid line) deteriorates from the true surface (dotted line) at larger distances from the starting condition, and the measured responses (points) start to decrease

10.3.3 Experimental Design

The first-order model with k factors has $k + 1$ parameters, and estimating these requires at least as many observations at different design points. Two options for an experimental design are shown in Fig. 10.3 for a two-factor design.

In both designs, we use several replicates at the starting condition as *center points*. This allows us to estimate the residual variance independently of any assumed model. Comparing this estimate with the observed discrepancies between any other point of the predicted response surface and its corresponding measured value allows testing the *goodness of fit* of our model. We also observe each factor at three levels, and can use this information to detect *curvature* not accounted for by the first-order model.

In the first design (Fig. 10.3A), we keep all factors at their starting condition level, and only change the level for one factor at a time. This yields two *axial points* for each factor, and the design resembles a coordinate system centered at the starting condition. It requires $2k + m$ measurements for k factors and m center point replicates and is adequate for a first-order model without interactions.

The second design (Fig. 10.3B) is a full 2^2-factorial where all *factorial points* are considered. This design would allow estimation of interactions between the factors and thus a more complex form of approximation. It requires $2^k + m$ measurements, but fractional replication reduces the experiment size for larger numbers of factors.

A practical problem is the choice for the low and high level of each factor. When chosen too close to the center point, the values for the response will be very similar and it is unlikely that we detect anything but large effects if the residual variance is not very small. When chosen too far apart, we might 'brush over' important features of the response surface and end up with a poor approximation. Only subject-matter

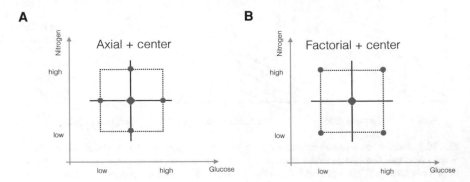

Fig. 10.3 Two designs for fitting a first-order RSM with two factors. **A** Center points and axial points that modify the level of one factor at a time only allow estimation of main effects, but not of interactions. **B** Center points and factorial points increase the experiment size for $k > 2$ but allow estimation of interactions

knowledge can guide us in choosing these levels satisfactorily; for biochemical experiments, one-half and double the starting concentration is often a reasonable first guess.[1]

10.3.4   Example—Yeast Medium Optimization

For the yeast medium optimization, we consider a 2^5-factorial augmented by six replicates of the center point. This design requires $2^5 + 6 = 38$ runs, which we reduce to $2^{5-1} + 6 = 22$ runs using a half-replicate of the factorial.

The design is given in Table 10.1, where low and high levels are encoded as ± 1 and the center level as 0 for each factor, irrespective of the actual amounts or concentrations. The last column shows the measured differences in optical density, with higher values indicating more cells in the corresponding well.

This encoding yields *standardized coordinates* with the center point at the origin $(0, 0, 0, 0, 0)$ and ensures that all parameters have the same interpretation as one-half the expected difference in response between low and high level, irrespective of the numerical values of the concentrations used in the experiment.

The model allows separation of three sources of variation: (i) the variation explained by the first-order model, (ii) the variation due to lack of fit, and (iii) the pure (sampling and measurement) error. These are quantified by an analysis of variance as shown in Table 10.2.

For example, the lack of fit is substantial, with a highly significant F-value, and the first-order approximation is clearly not an adequate description of the response surface. Still, the resulting parameter estimates already give a clear indication for the

---

[1] This requires that we use a log-scale for the levels of the treatment factors.

Table 10.1 Half-fraction of 2^5-factorial and center points. Variables are recoded from original levels to $-1/0/+1$. Last column: observed changes in optical density

Glc	N1	N2	Vit1	Vit2	Measured
Center point replicates					
0	0	0	0	0	81.00
0	0	0	0	0	84.08
0	0	0	0	0	77.79
0	0	0	0	0	82.45
0	0	0	0	0	82.33
0	0	0	0	0	79.06
Factorial points					
−1	−1	−1	−1	+1	35.68
+1	−1	−1	−1	−1	67.88
−1	+1	−1	−1	−1	27.08
+1	+1	−1	−1	+1	80.12
−1	−1	+1	−1	−1	143.39
+1	−1	+1	−1	+1	116.30
−1	+1	+1	−1	+1	216.65
+1	+1	+1	−1	−1	47.48
−1	−1	−1	+1	−1	41.35
+1	−1	−1	+1	+1	5.70
−1	+1	−1	+1	+1	84.87
+1	+1	−1	+1	−1	8.93
−1	−1	+1	+1	+1	117.48
+1	−1	+1	+1	−1	104.46
−1	+1	+1	+1	−1	157.82
+1	+1	+1	+1	+1	143.33

Table 10.2 ANOVA table for first-order response surface model. FO: first-order model in the given variables

	Df	Sum Sq	Mean Sq	F value	Pr(>F)
FO(Glc, N1, N2, Vit1, Vit2)	5	38102.72	7620.54	7.94	6.30e-04
Residuals	16	15355.32	959.71		
Lack of fit	11	15327.98	1393.45	254.85	3.85e-06
Pure error	5	27.34	5.47		

Table 10.3 Gradient of steepest ascent from first-order response surface model

Glc	N1	N2	Vit1	Vit2
−0.32	0.17	0.89	−0.09	0.26

Table 10.4 Parameter estimates of first-order response surface model

Parameter	Estimate	Std. error	t value	Pr(>|t|)
(Intercept)	85.69	6.60	12.97	6.59e-10
Glc	−15.63	7.74	−2.02	6.06e-02
N1	8.38	7.74	1.08	2.95e-01
N2	43.46	7.74	5.61	3.90e-05
Vit1	−4.41	7.74	−0.57	5.77e-01
Vit2	12.61	7.74	1.63	1.23e-01

next round of experimentation along the gradient. Recall that our main goal is to find the direction of highest increase in the response values and we are less interested in finding an accurate description of the surface in the beginning of the experiment.

The pure error is based solely on the six replicates of the center points and consequently has five degrees of freedom. Its variance is about 5.5 and very small compared to the other contributors. This means that the starting point produces highly replicable results and that small differences on the response surface are detectable.

The resulting gradient for this example is shown in Table 10.3, indicating the direction of steepest ascent in standardized coordinates.

The parameter estimates $\hat{\beta}_0, \ldots, \hat{\beta}_5$ are shown in Table 10.4. The intercept corresponds to the predicted average response at the center point; the empirical average is 81.12, in good agreement with the model.

We note that **N2** has the largest impact and should be increased, that **Vit2** should also be increased, that **Glc** should be decreased, while **N1** and **Vit1** seem to have only a comparatively small influence.

10.4 The Second-Order Model

10.4.1 Model

The second-order model adds purely quadratic (PQ) terms x_i^2 and two-way interaction (TWI) terms $x_i \cdot x_j$ to the first-order model (FO), such that the regression model equation becomes

$$y = f(x_1, \ldots, x_k) + e = \beta_0 + \sum_{i=1}^{k} \beta_i x_i + \sum_{i=1}^{k} \beta_{ii} x_i^2 + \sum_{i<j}^{k} \beta_{ij} x_i x_j + e \,,$$

and we again assume a constant error variance $\text{Var}(e) = \sigma^2$ for all points (and can again justify this because we are looking at a *local* approximation of the response surface). For example, the two-factor second-order response surface approximation is

$$y = \beta_0 + \beta_1 x_1 + \beta_2 x_2 + \beta_{1,1} x_1^2 + \beta_{2,2} x_2^2 + \beta_{1,2} x_1 x_2 + e \,.$$

and only requires estimation of six parameters to describe the true response surface locally. It performs satisfactorily for most problems, at least locally in the vicinity of a given point. Importantly, the model is still *linear in the parameters*, and parameter estimation can therefore be handled as a standard linear regression problem.

The second-order model allows curvature in all directions and interactions between factors which provide more information about the shape of the response surface. *Canonical analysis* then allows us to detect ridges along which two or more factors can be varied with only little change to the response and to determine if a stationary point on the surface is a maximum or minimum or a saddle-point where the response increases along one direction, and decreases along another direction. We do not pursue this technique and refer to the references in Sect. 10.6 for further details.

10.4.2 Path of Steepest Ascent

We find the direction of steepest ascent by calculating the gradient. In contrast to the first-order model, the gradient now depends on the position. The path of steepest ascent is then no longer a straight line, but is rather a curved path following the curvature of the response surface.

For example, the first component of the gradient (in direction x_1) depends on the levels $x_1, \ldots, x_k$ of all treatment factors in the design:

$$(g(x_1, \ldots, x_k))_1 = \frac{\partial}{\partial x_1} f(x_1, \ldots, x_k) = \beta_1 + 2\beta_{1,1} x_1 + \beta_{1,2} x_2 + \cdots + \beta_{1,k} x_k \,,$$

and likewise for the other $k - 1$ components.

10.4.3 Experimental Design

For a second-order model, we need at least three points on each axis to estimate its parameters and we need sufficient combinations of factor levels to estimate the interactions. An elegant way to achieve this is a *central composite design (CCD)*.

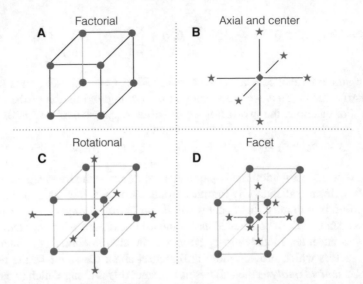

Fig. 10.4 A central composite design for three factors. **A** Design points for 2^3 full factorial. **B** Center point location and axial points chosen along axis parallel to coordinate axis and through center point. **C** Central composite design with axis points chosen for rotatability. **D** Combined design points with axial points chosen on facets introduce no new factor levels

It combines center points with replication to provide a model-independent estimate of the residual variance, axial points, and a (fractional) factorial design; Fig. 10.4 illustrates two designs and their components for a design with three factors.

With m center point replicates, a central composite design for k factors requires 2^k factorial points, $2 \cdot k$ axial points, and m center points. For our example with $k = 5$ factors, the full CCD with six replicates of the center points requires measuring the response for $32 + 10 + 6 = 48$ settings. The 2_V^{5-1} fractional factorial reduces this number to $16 + 10 + 6 = 32$ and provides a very economic design.

Choosing Axial Points

In the CCD, we combine axial points and factorial points, and there are two reasonable alternatives for choosing their levels: first, we can choose the axial and factorial points at the same distance from the center point, so they lie on a sphere around the center point (Fig. 10.4C). This *rotationally symmetric* design yields identical standard errors for all parameters and uses levels ± 1 for factorial and $\pm \sqrt{k}$ for axial points.

Second, we can choose axial points on the *facets* of the factorial hyper-cube by using the same low/high levels as the factorial points (Fig. 10.4D). This has the advantage that we do not introduce new factor levels, which can simplify the implementation of the experiment. The disadvantage is that the axial points are closer to the center points than the factorial points and estimates of parameters then have different standard errors, leading to different variances of the model's predictions depending on the direction.

Blocking a Central Composite Design

The factorial and axial parts of a central composite design are orthogonal to each other. This attractive feature allows a very simple blocking strategy, where we use one block for the factorial points and some of the replicates of the center point, and another block for the axial points and the remaining replicates of the center point. We can then conduct the CCD of our example in two experiments, one with $16 + 3 = 19$ measurements, and the other with $10 + 3 = 13$ measurements, without jeopardizing the proper estimation of parameters. The block size for the factorial part can be further reduced using the techniques for blocking factorials from Sect. 9.8.

In practice, this property provides considerable flexibility when conducting such an experiment. For example, we might first measure the axial and center points and estimate a first-order model from these data and determine the gradient.

With replicated center points, we are then able to quantify the lack of fit of the model and the curvature of the response surface. We might then decide to continue with the gradient pursuit based on the first-order model if curvature is small, or to conduct a second experiment with factorial and center points to augment the data for estimating a full second-order model.

Alternatively, we might start with the factorial points to determine a model with main effects and two-way interactions. Again, we can continue with these data alone, augment the design with axial points to estimate a second-order model, or augment the data with another fraction of the factorial design to disentangle confounded factors and gain higher precision of parameter estimates.

10.4.4 Sequential Experiments

Next, we measure the response for several experimental conditions along the path of steepest ascent. We then iterate the steps of approximation and gradient pursuit based on the condition with highest measured response along the path.

The overall sequential procedure is illustrated in Fig. 10.5: the initial second-order approximation (solid lines) of the response surface (dotted lines) is only valid locally (A) and predicts an optimum far from the actual optimum, but pursuit of the steepest ascent increases the response values and moves us closer to the optimum (B). Due to the local character of the approximation, predictions and measurements start to diverge further away from the starting point (C). The exploration around the new best condition predicts an optimum close to the true optimum (D) and following the steepest ascent (E) and a third exploration (F) achieves the desired optimization.

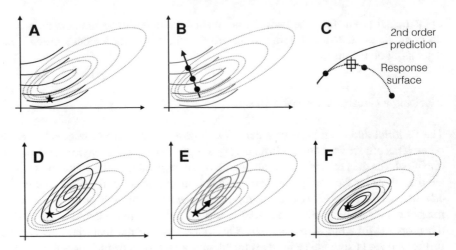

Fig. 10.5 Sequential experimentation for optimizing conditions using second-order approximations. **A** First approximation (solid lines) of true response surface (dotted lines) around a starting point (star). **B** Pursuing the path of steepest ascent yields a new best condition. **C** Slice along the path of steepest ascent. **D** Second approximation around new best condition. **E** Path of steepest ascent based on second approximation. **F** Third approximation captures the properties of the response surface around the optimum

The second-order model now has an optimum very close to the true optimum and correctly approximates the factor influences and their interactions locally.

A comparison of predicted and measured responses along the path supplies information about the range of factor levels for which the second-order model provides accurate predictions. A second-order model also allows prediction of the optimal condition directly. This prediction strongly depends on the approximation accuracy of the model, and we should treat predicted optima with suspicion if they are far outside the factor levels used for the exploration.

10.5   A Real-Life Example—Yeast Medium Optimization

We revisit our example of yeast medium optimization from Sect. 9.4. Recall that our goal is to alter the composition of glucose (**Glc**), monosodium glutamate (nitrogen source **N1**), a fixed composition of amino acids (nitrogen source **N2**), and two mixtures of trace elements and vitamins (**Vit1** and **Vit2**) to maximize growth of yeast. Growth is measured as increase in optical density, where higher increase indicates higher cell density. We summarize the four sequential experiments that are used to arrive at the specification of a new *high cell density (HCD) medium* (Roberts et al. 2020).

10.5.1   First Response Surface Exploration

Designing the response surface experiments required some consideration to account for the specific characteristics of the experimental setup. There are two factors that most influence the required effort and the likelihood of experimental errors: first, each experiment can be executed on a 48-well plate with cheap ingredients and any experiment size between 1 and 48 requires roughly equivalent effort. That means that we should use a second-order model directly rather than starting with a simpler model and augmenting the data if necessary. We decided to start from a defined yeast medium and use a 2^{5-1} fractional factorial design together with 10 axial points (26 design points without center points) as our first experimental design.

Second, it is very convenient to choose facet axial points, so each medium component has only three instead of five concentrations. We considered sacrificing the rotational symmetry of the design a small price to pay for easier and less error-prone pipetting.

We decided to use six center points to augment the five error degrees of freedom (26 observations in the design for 21 parameters in the second-order model). These allow separating lack of fit from residual variance. By looking at the six resulting growth curves, we could also gauge reproducibility of the initial experimental conditions.

We next decided to replicate the fractional factorial part of the CCD in the remaining 16 wells, using a biological replicate with new yeast culture and independent pipetting of the medium compositions. Indeed, the first replicate failed (cf. Sect. 9.4), and our analysis is based on the second replicate only. Using the remaining 16 wells to complete the 2^5-factorial would have been statistically desirable, but would require more complex pipetting and leave us without any direct replicates except the initial condition.

The resulting increases in optical density (OD) are shown in Fig. 10.6. The six values for the center points show excellent agreement, indicating high reproducibility of the experiment. Moreover, the average increase in OD for the starting condition is 81.1, and we already identified a condition with a value of 216.6, corresponding to a 2.7-fold increase.

We next estimated the parameters of the second-order approximation based on these observations; the estimates showed several highly significant and large main effects and two-way interactions, while the quadratic terms remained non-significant.

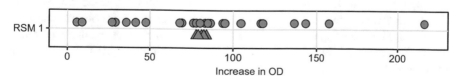

Fig. 10.6 Measured increase in OD for first exploration experiment. Triangles indicate center point replicates

Table 10.5 ANOVA table for second-order response surface approximation

	Df	Sum Sq	Mean Sq	F value	Pr(>F)
FO(Glc, N1, N2, Vit1, Vit2)	5	43421.11	8684.22	99.14	9.10e-09
TWI(Glc, N1, N2, Vit1, Vit2)	10	15155.26	1515.53	17.3	2.49e-05
PQ(Glc, N1, N2, Vit1, Vit2)	5	289.44	57.89	0.66	6.61e-01
Residuals	11	963.6	87.6		
Lack of fit	6	936.26	156.04	28.54	1.02e-03
Pure error	5	27.34	5.47		

Table 10.6 Predicted stationary point of first approximation in standardized coordinates

Glc	N1	N2	Vit1	Vit2
4.77	0.38	−2.37	2.19	1.27

This shows that optimization cannot be done iteratively one factor at a time, but that several factors need to be changed simultaneously. The non-significant quadratic terms indicate that either the surface is sufficiently plane around the starting point or that we chose factor levels too close together or too far apart to detect curvature. The ANOVA table for this model (Table 10.5), however, shows that quadratic terms and two-way interactions as a whole cannot be neglected, but that the first-order part of the model is by far the most important.

The table also suggests considerable lack of fit, but we expect this at the beginning of the optimization. This becomes very obvious when looking at the predicted stationary point of the approximated surface (Table 10.6), which contains negative concentrations and is far outside the region we explored in the first experiment (note that the units are given in standardized coordinates, so a value of 10 means 10 times farther from the center point than our axial points).

10.5.2   First Path of Steepest Ascent

Based on this second-order model, we calculated the path of steepest ascent for the second experiment in Table 10.7.

The first column shows the distance from the center point in standardized coordinates, so this path moves up to 4.5 times further than our axial points. The next columns are the resulting levels for our five treatment factors in standardized coordinates. The column 'Predicted' provides the predicted increase in OD based on our second-order approximation.

Table 10.7 Path of steepest ascent in standardized coordinates, predicted increase in OD, and measured increase in OD based on first exploration data

Distance	Glc	N1	N2	Vit1	Vit2	Predicted	Measured
0.0	0.0	0.0	0.0	0.0	0.0	81.8	78.3
0.5	−0.2	0.1	0.4	0.0	0.1	108.0	104.2
1.0	−0.4	0.3	0.8	0.0	0.3	138.6	141.5
1.5	−0.6	0.6	1.1	0.0	0.5	174.3	156.9
2.0	−0.8	0.8	1.4	0.1	0.8	215.8	192.8
2.5	−1.1	1.1	1.7	0.2	1.0	263.3	238.4
3.0	−1.3	1.4	1.9	0.3	1.3	317.2	210.5
3.5	−1.5	1.7	2.1	0.4	1.5	377.4	186.5
4.0	−1.7	2.1	2.4	0.5	1.8	444.3	144.2
4.5	−1.9	2.4	2.6	0.6	2.0	517.6	14.1

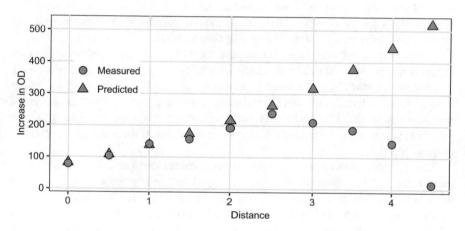

Fig. 10.7 Observed growth and predicted values along the path of steepest ascent. Model predictions deteriorate from a distance of about 2.5–3.0

We experimentally followed this path and took measurements at the indicated medium compositions. This resulted in the measurements given in column 'Measured'. We observe a fairly good agreement between prediction and measurement up to about distance 2.5–3.0. Thereafter, the predictions keep increasing while measured values drop sharply to almost zero. This is an example of the situation depicted in Fig. 10.5C, where the approximation breaks down further from the center point. This is clearly visible when graphing the prediction and measurements in Fig. 10.7. Nevertheless, the experimental results show an increase by about threefold compared to the starting condition along the path.

10.5.3 Second Response Surface Exploration

In the next iteration, we explored the response surface locally using a new second-order model centered at the current best condition. We proceeded just as for the first RSM estimation, with a 2^{5-1}-design combined with axial points and six center point replicates.

The 32 observations from this experiment are shown in Fig. 10.8 (top row). The new center conditions have consistently higher values than the conditions of the first exploration (bottom row), which is a good indication that the predicted increase is indeed happening in reality. Several conditions show further increase in response, although the fold-changes in growth are considerably smaller than before. The maximal increase from the center point to any condition is about 1.5-fold. This is expected when we assume that we approach the optimum condition.

The center points are more widely dispersed now, which might be due to more difficult pipetting, especially as we originally used a standard medium. The response surface might also show more curvature in the current region, such that deviations from a condition result in larger changes in the response.

Importantly, we do not have to compare the observed values between the two experiments directly, but need only compare a fold-change in each experiment over its center point condition. That means that changes in the measurement scale (which is in arbitrary units) do not affect our conclusions. Here, the observed center point responses of around 230 agree very well with those predicted from the gradient pursuit experiment, so we have confidence that the scales of the two experiments are comparable.

The estimated parameters for a second-order model now present a much cleaner picture: we find only main effects large and significant, and the interaction of glucose with the second nitrogen source (amino acids); all other two-way interactions and the quadratic terms are small. The ANOVA Table 10.8 still indicates some contributions of interactions and purely quadratic terms, but the lack of fit is now negligible. In essence, we are now dealing with a first-order model with one substantial interaction.

The model is still a very local approximation, as is evident from the stationary point far outside the explored region with some values corresponding to negative concentrations (Table 10.9).

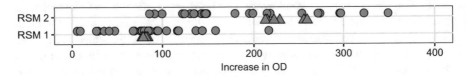

Fig. 10.8 Distribution of measured values for first (bottom) and second (top) exploration experiment with 32 conditions each. Triangles indicate center point replicates

Table 10.8 ANOVA table for second-order response surface approximation

	Df	Sum Sq	Mean Sq	F value	Pr(>F)
FO(Glc, N1, N2, Vit1, Vit2)	5	89755.57	17951.11	24.43	1.31e-05
TWI(Glc, N1, N2, Vit1, Vit2)	10	33676.72	3367.67	4.58	9.65e-03
PQ(Glc, N1, N2, Vit1, Vit2)	5	28320.87	5664.17	7.71	2.45e-03
Residuals	11	8083.92	734.9		
Lack of fit	6	5977.97	996.33	2.37	1.82e-01
Pure error	5	2105.96	421.19		

Table 10.9 Predicted stationary point of second exploration in standardized coordinates

Glc	N1	N2	Vit1	Vit2
1.02	−6.94	1.25	−11.76	−2.54

10.5.4 Second Path of Steepest Ascent

Based on this second model, we again calculated the path of steepest ascent and followed it experimentally. The predictions and observed values in Fig. 10.9 show excellent agreement, even though the observed values are consistently higher. The systematic shift already occurs for the starting condition and a likely explanation is the gap of several months between second exploration experiment and the gradient pursuit experiment, leading to differently calibrated measurement scales.

This systematic shift is not a problem in practice, since we have the same center point condition as a baseline measurement, and in addition are interested in changes in the response along the path, and less in their actual values. Along the path, the

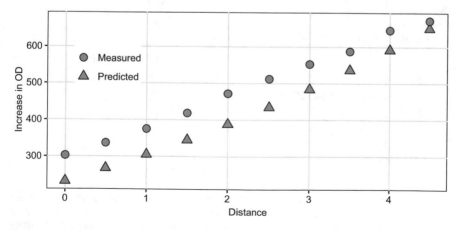

Fig. 10.9 Observed growth and predicted values along the second path of steepest ascent

response is monotonically increasing, resulting in a further 2.2-fold increase compared to our previous optimum at the last condition.

Although our model predicts that further increase in growth is possible along the gradient, we stopped the experiment at this point. The wells of the 48-well plate were now crowded with cells, and several components were so highly concentrated that salts started to fall out from the solution at the distance-5.0 medium composition.

10.5.5   Conclusion

Overall, we achieved a threefold increase in the first iteration, followed by another increase of 2.2-fold in the second iteration, yielding an overall increase in growth of about 6.8-fold. Given that previous manual optimization resulted in a less than three-fold increase overall, this highlights the power of experimental design and systematic optimization of the medium composition.

We summarize the data of all four experiments in Fig. 10.10. The vertical axis shows the four sequential experiments and the horizontal axis the increase in OD

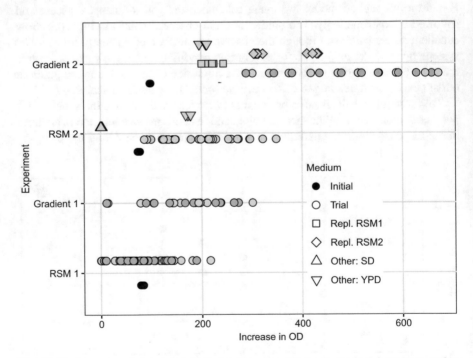

Fig. 10.10 Summary of data acquired during four experiments in sequential response surface optimization. Additional conditions were tested during second exploration and gradient pursuit (point shape), and two replicates were measured for both gradients (point shading). Initial starting medium is shown in black

for each tested condition. Round points indicate trial conditions of the experimental design for RSM exploration or gradient pursuit, with grey shading denoting two independent replicates. During the second iteration, additional standard media (YPD and SD) were also measured to provide a direct comparison against established alternatives. The second gradient pursuit experiment repeated previous exploration points for direct comparison.

We noted that while increase in OD was excellent, the growth rate was much slower compared to the starting medium composition. We addressed this problem by changing our optimality criterion from the simple increase in OD to an increase penalized by duration. Using the data from the second response surface approximation, we calculated a new path of steepest ascent based on this modified criterion and arrived at the final high-cell density medium which provides the same increase in growth with rates comparable to the initial medium (Roberts et al. 2020).

10.6 Notes and Summary

Notes

Sequential experimentation for optimizing a response was already discussed in Hotelling (1941), and the response surface methodology was introduced and popularized for engineering statistics by George Box and co-workers (Box and Wilson 1951; Box and Hunter 1957; Box 1954); a current review is Khuri and Mukhopadhyay (2010). Two relevant textbooks are the introductory account in Box et al. (2005) and the more specialized text by Box and Draper (2007); both also discuss canonical analysis. The classic Cochran and Cox (1957) also provides a good overview. The use of RSM in the context of biostatistics is reviewed in Mead and Pike (1975).

Using R

The rsm package (Lenth 2009) provides the function rsm() to estimate response surface models; they are specified using R's formula framework extended by convenience functions (Table 10.10). Coding of data into $-1/0/+1$ coordinates is done using coded.data(), and steepest() predicts the path of steepest ascent. Central composite designs (with blocking) are generated by the ccd() function.

Table 10.10 Shortcuts for defining response surface models in rsm()

Shortcut	Meaning	Example for two factors
FO()	First-order	$\beta_0 + \beta_1 x_1 + \beta_2 x_2$
PQ()	Pure quadratic	$\beta_{1,1} x_1^2 + \beta_{2,2} x_2^2$
TWI()	Two-way interactions	$\beta_{1,2} x_1 x_2$
SO()	Second-order	$\beta_0 + \beta_1 x_1 + \beta_2 x_2 +$ $\beta_{1,2} x_1 x_2 + \beta_{1,1} x_1^2 + \beta_{2,2} x_2^2$

Summary

Response surface methods provide a principled way for finding optimal experimental conditions that maximize (or minimize) the measured response. The main idea is to create an experimental design for estimating a local approximation of the response surface around a given point, determine the path of steepest ascent using the approximation model and then experimentally explore this path. If a 'better' experimental condition is found, the process is repeated from this point until a satisfactory condition is found.

A commonly used design for estimating the approximation model is the central composite design. It consists of a (fractional) factorial design augmented by axial points. Conveniently, axial points and the factorial points are orthogonal and this allows us to implement the design in several stages, where individual smaller experiments are run for the axial points and for (parts of the fractional) factorial points. Multiple replicates of the center point allow separation of lack of fit of the model from residual variance.

References

Box, G. E. P. (1954). "The Exploration and Exploitation of Response Surfaces: Some General Considerations and Examples". In: Biometrics 10.1, p. 16.

Box, G. E. P. and N. R. Draper (2007). Response Surfaces, Mixtures, and Ridge Analyse. Wiley & Sons, Inc.

Box, G. E. P. and J. S. Hunter (1957). "Multi-Factor Experimental Designs for Exploring Response Surfaces". In: The Annals of Mathematical Statistics 28.1, pp. 195–241.

Box, G. E. P. and K. B. Wilson (1951). "On the Experimental Attainment of Optimum Conditions". In: Journal of the Royal Statistical Society Series B (Methodological) 13.1, pp. 1–45.

Box, G. E. P., J. S. Hunter, and W. G. Hunter (2005). Statistics for Experimenters. Wiley, New York.

Cochran, W. G. and G. M. Cox (1957). Experimental Designs. John Wiley & Sons, Inc.

Hotelling, H. (1941). "Experimental Determination of the Maximum of a Function". In: The Annals of Mathematical Statistics 12.1, pp. 20–45.

Khuri, A. I. and S. Mukhopadhyay (2010). "Response surface methodology". In: WIREs Computational Statistics 2, pp. 128–149.

Lenth, R. V. (2009). "Response-surface methods in R, using RSM". In: Journal of Statistical Software 32.7, pp. 1–17.

Mead, R. and D. J. Pike (1975). "A Biometrics Invited Paper. A Review of Response Surface Methodology from a Biometric Viewpoint". In: Biometrics 31.4, pp. 803–851.

Roberts, T. M., H.-M. Kaltenbach, and F. Rudolf (2020). "Development and optimisation of a defined high cell density yeast medium". In: Yeast 37 (5-6), pp. 336–347.

Index

© Springer Nature Switzerland AG 2021
H.-M. Kaltenbach, *Statistical Design and Analysis of Biological Experiments*,
Statistics for Biology and Health, https://doi.org/10.1007/978-3-030-69641-2

Printed in the United States
by Baker & Taylor Publisher Services